PRAISE FOR *THE POSSIBILITY PRINCIPLE*

"This book brilliantly applies the principles of quantum physics to our everyday life in a very personal and healing way. Mel shows us how to develop authentic self-esteem and thrive in our relationships by embracing this new worldview."

CAROLINE MYSS
author of *Anatomy of the Spirit* and *Defy Gravity*

"Mel Schwartz draws breathtaking parallels between the quantum reality and our human experience, revealing powerful lessons for living to our greatest potential. He shows that consciousness is fundamental and that our minds and hearts are shared as one. The implications for human welfare are profound."

LARRY DOSSEY, MD
author of *One Mind: How Our Individual Mind Is Part of a Greater Consciousness and Why It Matters*

"This handbook for life brings the inspiration of quantum uncertainty, interconnectedness, emergence, and entanglement into compelling practices to help each of us grow and heal. A seasoned clinician, Schwartz offers new perspectives on the power of our beliefs, worldviews, and intentions that can help us enter realms of unfolding possibilities."

MARILYN SCHLITZ, PHD
professor and chair of PhD programs in
transpersonal psychology, Sofia University; author of
the book *Death Makes Life Possible*; executive producer
and writer of the film *Death Makes Life Possible*

"*The Possibility Principle* clearly describes the havoc wrought by a worldview that reinforces competition and separateness while neglecting compassion and cooperation. Schwartz . . . calls for nothing less than new ways of thinking that will turn around both individual lives and the world at large. This move from a deterministic, mechanistic worldview to one that is participatory has incredible implications for our personal lives, for our mental and emotional health, and for the very foundations of society."

STANLEY KRIPPNER, PHD
coauthor of *Personal Mythology*

"*The Possibility Principle* is a beautifully written book on how we can live fully as we awaken from the stupor of Newton's sleep. The benefits range from overcoming anxiety and depression to mastering our communication to thriving in our relationships!"

ALLAN COMBS, PHD
author of *The Radiance of Being* and *Consciousness Explained Better*

THE
POSSIBILITY
PRINCIPLE

HOW QUANTUM PHYSICS
CAN IMPROVE THE WAY YOU
THINK, LIVE, AND LOVE

THE
POSSIBILITY
PRINCIPLE

MEL SCHWARTZ

sounds true
BOULDER, COLORADO

Sounds True
Boulder, CO 80306

This book is not intended as a substitute for the medical recommendations of
physicians, mental health professionals, or other health-care providers. Rather, it is
intended to offer information to help the reader cooperate with physicians, mental
health professionals, and health-care providers in a mutual quest for optimal well-
being. We advise readers to carefully review and understand the ideas presented and
to seek the advice of a qualified professional before attempting to use them.

Some names and identifying details have been changed to protect the privacy of
individuals.

Published 2017

Cover design by Jennifer Miles
Book design by Beth Skelley

Printed in Canada

Library of Congress Cataloging-in-Publication Data
Names: Schwartz, Mel, author.
Title: The possibility principle : how quantum physics can improve the way
    you think, live, and love / Mel Schwartz.
Description: Boulder, CO : Sounds True, 2017. | Includes bibliographical
    references and index.
Identifiers: LCCN 2016053568| ISBN 9781622038633 (hardcover) |
    ISBN 9781622038640 (ebook)
Subjects: LCSH: Science and psychology. | Quantum theory.
Classification: LCC BF64 .S39 2017 | DDC 530.12—dc23
LC record available at https://lccn.loc.gov/2016053568

10 9 8 7 6 5 4 3 2 1

For my partner, Leslie; my sons, Alex and Jesse;
and for my departed parents, Ruth and Sidney,
with love and gratitude.

CONTENTS

INTRODUCTION

I awakened one weekend morning some twenty years ago feeling quite anxious. Having recently divorced, I was missing my kids, who were with their mom that weekend, and I thought getting out of the house might help. I headed out for a bike ride, but as I pedaled around the gently rolling hills of my hometown, my angst only continued to heighten. I realized I was experiencing the start of an anxiety attack—something that had never happened to me before.

Still feeling shaken when I arrived home, I walked into my office and absently pulled a book off the shelf: *The Turning Point: Science, Society, and the Rising Culture*, written by theoretical quantum physicist Fritjof Capra, which I had bought some time before but never gotten around to reading. Now I opened it and began to read about a major worldview shift, catalyzed by quantum physics (also known as quantum mechanics), which was just beginning to herald a deep impact on many aspects of our society. Capra described how this quantum sense of reality departed radically from our beliefs rooted in the once-groundbreaking work of the seventeenth-century thinkers Isaac Newton and René Descartes.[1] I felt myself becoming enthralled as I read of a wondrous universe—one that was inseparably whole and thoroughly interpenetrating—in which all notions of separation faded away. Moreover, this new reality indicated that the universe existed in a state of uncertainty—a state of pure potentiality.

As I continued to read, I was astonished to notice that my anxiety and despair had been supplanted by a sense of serenity and connectedness. Although nothing at all had changed in my outer world, my inner world was undergoing a profound shift. I was becoming a participant in this magical wholeness that I was reading about. I no longer felt alone, but sensed that I was an integral part of this universe. I felt calm and connected. Capra's vision had opened me to the immense potential to be derived from connectedness.

As I continued reading over the following days, my fears retreated, and I began to embrace my future with confidence that I could summon these new potentials into being. Finishing Capra's book, I set to reading more about quantum physics, focusing on the theory and its implications (as the mathematical and technical aspects were well beyond my comprehension). I had immersed myself not only in understanding the science behind this quantum vision but also, more importantly, how it might affect us on personal levels. As my understanding of reality altered to align with these discoveries, my beliefs changed too. I reflected on how my misinformed beliefs, rooted in fear and the idea that change was onerous, had constrained my life. This insight ultimately touched virtually every aspect of how I think and live.

The new life that emerged was untethered by many of my old limitations. Rather than feeling unmoored, I marveled at my transformative experience, reflecting that if this worked so well for me, why not for others?

I began to integrate these insights into my work as a psychotherapist and marriage counselor and was further buoyed by the success experienced by so many of my clients. Over time, I developed an accessible approach that applied the quantum worldview to personal growth, showing people how to live more resiliently and fearlessly and how to think, feel, relate, and communicate differently based on this worldview.

At the same time, I began to teach this approach through a series of lectures and workshops to both therapists and the general public. This experience of working with so many people and witnessing their gains deepened my conviction about this approach toward personal transformation. This process helped me distill my new thinking into some basic principles to assist people in their lives.

How exactly do the quantum worldview and its core principles bring about personal transformation? It boils down to one word:

possibility. I've called this book *The Possibility Principle* because its purpose is to open the window of possibilities in all areas of our lives. I believe that we can shed the old beliefs, thoughts, and behaviors that have constrained us by welcoming life-enhancing principles that we can cull from quantum physics. Doing so enables our proactive participation in orchestrating our life experiences.

This book shares the myriad benefits we can enjoy by integrating the messages of quantum physics into our everyday existence. Examples drawn from my daily practice will show you how you too can achieve these breakthroughs. (I have altered the names and identities of all clients discussed in this book to preserve confidentiality.)

When we embrace the possibilities the quantum worldview offers us, we invite defining moments—moments when we dare to venture into new terrain, transcending our struggles and actualizing new realities. For example, reading Capra's *The Turning Point* was a defining moment for me. Defining moments are far more than simple insights. Regrettably, most of our insights—our *aha* moments—fade over time because, mired in the comfort of the familiar, we don't treat them with the respect they deserve. A defining moment is a singular burst of insight in which we choose to embark in a new direction. All that is required to break new ground is our willful intention to do so. At the core of the Possibility Principle is this truth: possibility begets more and more possibility.

Many books and teachings about personal growth address our thinking and perhaps our beliefs. Yet they leave out something critically important: an appreciation of how our operating worldview informs our beliefs, our thinking, and our lives. I'll demonstrate how our operating worldview creates the landscape we live in and, as a result, is the basis of our experience. Because I'm not a scientist, my descriptions of quantum physics are easy to understand and intended as metaphorical, not literal. That said, quantum physics resists the imposition of stark definitions even by experts in the field.

By helping my clients work through a vast array of challenges, I have developed practices that can help us overcome our obstacles and summon new possibilities into our lives. Many of our operating beliefs, when we look deeply into them, make little sense and do much harm. Throughout this book, I introduce exercises—prompts for self-inquiry—that will help you reformulate your worldview and apply quantum principles to your thinking and beliefs. Mastering your thinking and beliefs will lay the foundation for authentic self-esteem, help you overcome chronic, cumbersome issues, and improve your ability to communicate. All of these shifts in turn can open you to new possibilities in your relationships with others, as well as your relationship with yourself.

I begin the book by describing the transformation in scientific understanding that has taken place over the past three centuries, how it has shifted from the mechanistic model of the universe conceived by thinkers such as Newton and Descartes to the revolutionary insights of quantum physics developed in the last ninety years. I go on to show how our unconscious addiction to the old worldview damages our ability to live sensibly, let alone to achieve our greatest potential. Each of the next three chapters explores a key principle in quantum physics that we can utilize in personal ways to our great benefit. In chapters 5–12, I present many examples to demonstrate how to make use of basic quantum concepts to reconstruct how we think about ourselves, how we interact with people and our environment, and how we communicate with each other.

The goal of this book is to help you reenvision your world and yourself as you develop the insights and skills to navigate your life without fear and with self-empowerment. You'll learn methods to become the master of your thinking and discover profound meaning and purpose in your life.

The time has come for each of us to experience a personal revolution, just as the scientific world did nine decades ago. It is time for us to let go of a long-outdated worldview and embrace a new, forward-thinking, empowering worldview offered by quantum physics. It is time for each of us to experience new possibilities—possibilities we can apprehend when we let go of the old outlook that keeps us stuck in so many areas of our lives. Reading this book is your first step. Welcome to your new quantum life.

THE PROMISE OF
THE QUANTUM WORLDVIEW

A worldview is the "meta-picture" of how we think reality operates. What we need to understand is that it's merely a temporary snapshot of reality. Over time, new theories and discoveries presage shifts in our worldviews. The paradigm that Earth was the center of the universe prevailed until it was eventually overturned in the early 1600s by Galileo, who suffered persecution for claiming that Earth moved around the sun. What we think of as reality is actually no more than the current worldview. Yet as each new theory and discovery arises, we reenvision the world and how we come to live in it, and our new vision has profound consequences on most aspects of our lives.

Until the sixteenth century, the worldview in the West was dominated by the teachings of the Roman Catholic Church, which were based in part on the writings of Aristotle, amounting to a combination of reason and faith. Then a series of developments in physics and astronomy, brought about by early scientific masterminds Nicolaus Copernicus, Johannes Kepler, and Galileo Galilei, led to a greater reliance on analytical reasoning and a view of the natural world based on mathematics rather than spiritual principles or superstition. The philosopher and mathematician René Descartes followed the astronomical discoveries of these men—including the heliocentric model of the solar system—and discovered what he believed to be the absolute certainty and truth of scientific knowledge. Among other things, he described the universe as operating like a giant clock.[1]

Isaac Newton straddled the seventeenth and eighteenth centuries and furthered Descartes's vision by depicting the universe as a giant machine, a worldview known as the *mechanistic paradigm*. His vision dominated science and philosophy until the twentieth century, when quantum physics revealed that reality on the quantum level—dealing with the smallest ingredients of the universe but having implications for our everyday world as well—looked strikingly different from what we could have imagined.

It is my impassioned conviction—reinforced through both my professional and personal experience—that nothing impedes our lives and our ability to thrive as distinctly as the outdated, seventeenth-century worldview of *mechanism*. This outmoded paradigm imposes a straitjacket on our beliefs and our thinking, impoverishing our life experiences. Not having kept pace with the advances of science and its ensuing shift in philosophical thinking, many of our operating assumptions and beliefs are obsolete, incoherent, and invalid. Not only has mechanism fostered a disenchanted view of the cosmos, and consequently our lives, but it assaults our individual and collective psyches as well.

The mechanistic model of the world says that the world consists of separate and inert objects that are disconnected from one another, interacting only through cause and effect. According to this picture of reality, the world operates as a giant machine, and we become the cogs in the machine, detached from one another and disconnected from the universe at large. Separation being the essence of the Newtonian worldview, this image is devoid of any sense of relatedness, leaving us as strangers in a cold, austere world, absent any scintilla of belonging or purpose.

Through this filter, we experience a vast array of struggle and malaise: anxiety, depression, failed relationships, incoherent communication, and the gloom of existential despair—all of which we will explore later in detail. The motif of separation isolates us and induces us to compete rather than collaborate and promotes extreme individualism over the common good. Winning replaces compassion. Conflict trumps cooperation. This crippling worldview colonizes our thoughts and beliefs, and for the most part, we live out our lives in accordance with this construct.

Another core tenet of the mechanistic paradigm is known as *determinism*—the ability to predict future conditions based on present circumstances. Determinism strips us of our sense of wonder and creativity, but even worse, it drives us to seek certainty as we avoid the uncertain. As a culture, the epidemic of anxiety that we experience is caused in large part by our addiction to certainty, which has us fear and avoid the unknown. Certainty also imperils our relationships as it thwarts our ability to be truly present.

From this cosmology we succumb to the dispiriting mechanization of our spirit. Meaning and purpose are cast out in deference to cause and effect—the fundamental by-product of separation. It is little wonder that we suffer as we do. As the noted eco-philosopher Henryk Skolimowski succinctly put it, "As we read the universe, so we act in it."[2] If we constantly envision a machine operating, we become machinelike. If we see and think in separation, we'll experience our lives through this lens of isolation. Imagine wearing very dark sunglasses permanently affixed to your face. You wouldn't see things the same way as others would. The tint of your glasses would filter the light. The worldview of mechanism is the dark filter through which most of us see.

We need to examine how our prevailing worldview corresponds with its consequences. We are clearly operating from the wrong game plan. Science has indicated—and empirically proven, at least for the time being—that its new discoveries demand a significant reconsideration of our worldview. Yet most of us remain wed to the old principles that classical science postulated, and our lives are terribly impoverished as a result.

To use another metaphor, imagine that you're a fish in a fishbowl. Your universe, your reality, is limited to the edges of the bowl that you keep swimming into. Similarly, our reality is confined by mechanistic determinism and separation. These false beliefs do unimaginable harm to us, as we'll explore together.

RETHINKING REALITY

The remarkable discoveries emerging from the field of quantum physics over the last century have been discussed by scientists and noted by philosophers, but we haven't succeeded in adopting them in our everyday lives. That's because most people aren't sufficiently aware of these startling breakthroughs or—more importantly—of the beneficial implications for how we live and view reality.

The primary principles from quantum physics show up in three vital conceptions that can enable us to live the lives we choose.

1. **Embrace uncertainty.**

   Contrary to Newton's determinism, one of the core findings of the new, quantum-based science is that the universe is awash in uncertainty. In the early part of the twentieth century, the physicist Werner Heisenberg discovered that within the quantum realm, the rule of certainty no longer prevailed. Uncertainty is the fabric of the quantum world. This principle has vast applications in our lives, and we should view it as good news. Welcoming uncertainty frees us from the severely constrained existence in which the mechanistic template imprisons us. Think of uncertainty as the wind in our sails, empowering us toward the lives we seek. Uncertainty is where new possibility lies.

2. **The universe is in a pure state of potential.**

   Uncertainty implies potentiality, as all outcomes are possible. It appears that reality looks more like a *reality-making process*—a perpetual state of flow—than a fixed state of being. With a shift in our perceptions and thinking, we too can enter into the flow of possibilities. We are no longer inert cogs in the giant machine but the creators of our own destinies.

3. **The universe appears fundamentally inseparable.**
   The universe appears to be a thoroughly interconnected, interpenetrating whole. Inseparability implies that we are an integral part of everything and everyone. This understanding showers us with meaning, purpose, and connectedness. Once we appreciate that the distinction between the other and ourselves is altogether indistinct, inseparability can usher in compassion and empathy. As we'll see, inseparability becomes the bedrock of healthy relationships.

These three principles stand in stark opposition to how and what we think reality looks like. You may wonder what these principles of quantum physics have to do with your personal life; the answer is *virtually everything*. This includes the possibility of living a self-empowered and meaningful life as we replace the tired beliefs that constrict and dehumanize us and welcome the profound benefits to be gained by opening to the messages of quantum physics. In the next three chapters we'll explore each of these principles in some detail, and throughout the rest of book, we'll unpack their beneficial effects in greater depth.

Given that the universe appears to be inexorably flowing, and that we are an inextricable part of that reality, quantum physics provides a figurative ride on the current of change. We cease being the inert parts of Newton's machine, limited by determinism, and become conscious participants in the crafting of our lives. From this perpetual condition of movement and flux, we are left not with a fixed material reality but a bubbling state of potentiality.

Yet if we keep selecting the same habitual thoughts and feelings, we remain stuck. This common experience is well illustrated in the 1993 movie *Groundhog Day*. The main character, played by Bill Murray, is given repeated opportunities to remake himself by reliving the same day over and over again. In what he at first experiences as a feverish nightmare, he is allowed to return continually to the previous day and learn from his mistakes and those of others around him. Once he adapts to what seems like a punishment or a maddening existential trap, he takes the opportunity to choose differently and summon new possibilities, freeing himself from his habitual repetition of the past. In the process, he becomes more aware of his interconnectedness with the people in his life and, as a result, a more compassionate person.

Most of us struggle in a similar way to achieve such change in our lives. The quantum model invites us into our growth and change process; this new perception of reality enables us to apprehend the possibilities that await us. The principles of quantum physics enable us to break free from the past as we may choose. Uncertainty, potentiality, and inseparability provide us with the platform to become the masters of our lives. If we are all interconnected in an unimaginably profound way, and the universe is uncertain and constantly in flux, we are liberated from the confines of predictability, and we open the door to new potential and personal growth. Once we grasp the fact that consciousness, not matter, appears to be the fundamental basis of reality, we are in the director's chair.

This whole new vision of reality created by the discoveries of quantum physics is also known as the *participatory worldview*. The revelation of the participatory worldview is that reality appears to be a kind of creative dance in which we all participate—again, more of a reality-making process than a fixed, objective reality.

To continue to adhere to the classical—mechanistic—paradigm, which reduces us to meaningless, disconnected parts of the giant machine, is akin to living out our lives in solitary confinement, falsely imprisoned by our operating beliefs. Our departure from our old mind-set enables us to align with the flowing potentiality of the universe. We become unstuck. Opening to the messages from quantum physics restores our human potential. When we alter our view of reality from a lifeless, inhospitable machine to a wondrous, inseparable universe of possibility, everything changes. With this shift of mind, our possibilities become bountiful.

2

WHY WE NEED TO
EMBRACE UNCERTAINTY

Many of us are enraptured by spectator sports because of the thrill of not knowing the outcome. After all, how exciting would it be to watch the replay of the game if we already knew who won? We may read an intriguing book or go to a suspenseful movie to seek the excitement of the unknown. Mysteries and thrillers captivate us because we don't know how they'll turn out. Some people play games of chance or the lottery because while they hope to win, they cannot be certain that they will. Our uncertainty is what engages us and, at times, makes us feel spellbound. Uncertainty drives a substantial percentage of the entertainment industry and the US gross national product. The uncertain presents limitless possibilities and attracts our full attention. It allows us to be present and enlivens our lives.

Yet here's the paradox: We often seek these exciting, uncertain experiences as compensation for living lives dulled by our need for certainty and predictability—a need that tends to rule our thinking and our decision-making. When we become addicted to this need, we live out our lives in a formatted way, less present and mindful than we could be. Being bound up in the straitjacket of certainty makes us like a character in a novel for whom the plot is already written. We're simply living according to our script.

As explained in chapter 1, uncertainty is one of the keystones of the new quantum science and directly challenges our old worldview that has us value certainty and predictability. The strong appeal of uncertainty in sports and entertainment should be a clue to how much we might gain if we reconsider our entire relationship with certainty. Over the years of my working with people in therapy and facilitating workshops, it has become clear to me that our attachment to certainty can have debilitating consequences. I have encountered so many people whose struggles with anxiety and depression are caused by being overly wed to certainty.

HOW THE NEED FOR CERTAINTY PROVOKES ANXIETY

Tom came to see me seeking help for his self-described anxiety disorder. He held down an executive-level job and supported his family well, yet he had no peace of mind, let alone joy. He was deluged by his fear of not knowing what the future held. His need for certainty overwhelmed him. As he was preparing for a presentation at work, he became distressed by the uncertainty that lay ahead of him: How would his talk be received? What questions might be asked of him? Was he sufficiently prepared? Tom's worries had a negative impact on his work, as the very thing he feared—a mediocre performance at his job, accompanied by a gloomy aspect that turned off coworkers—was precisely what he was creating. At home, Tom questioned whether his wife would continue to love him in the future, turning her off as she understandably pulled back from him. In essence, Tom was addicted to seeking the assurance that certainty might provide.

I've come to see that anxiety disorders are often correlated to our demand for certainty: the greater our dependence on predictability, the more we experience anxiety. What Tom couldn't see was that this level of certainty and predictability is not only elusive, but also in most cases undesirable. To live in such a way precludes spontaneity or creativity, which require your mind to be present in the moment.

Tom's troubles, like countless other people's, were unconsciously informed by his operating worldview. As I've said, the roots of our dependence on certainty go back more than three hundred years, to Isaac Newton's seventeenth-century worldview and its core pillar of "determinism." Newton saw reality as composed primarily of objects, and all objects, including humans, as separate and distinct from one another. This worldview proposed that with sufficient present measurements we could determine the future. Think of a billiard ball striking another ball and causing it to carom off the side of the pool table. With precise calculation, we can predict where the ball will make contact with the next object—be it a side-cushion, a pocket, or another ball—and plot the entire unfolding sequence of events.

This construct sets up an arrow from past to present to future, imposing a cause-and-effect view of reality through which present conditions order the circumstances of the future. As a result, we lose our autonomy, and our human consciousness is subverted to being no more than a billiard ball struck by a cue stick. Tom's thoughts were the equivalent of the ball, following its predictable path in an attempt to ferret out the future.

Romance, like sports and mysteries, is another area of life where we can easily see how certainty imprisons us. The wondrous feeling of falling in love is an immersion into the unknown. It's rich in adventure and discovery and the uncertainty that underscores the sense of Eros: "I feel like I love her, but I'm not sure if she loves me." So the uncertainty has me hang on her every word and facial expression.

But once we secure the relationship, romance tends to wither as we become less present to each other. We think we know in advance how our partners will act or respond; we finish their thoughts for them (mentally and sometimes verbally) even before they conclude their sentences. As a result, sexual intimacy loses vitality and passion with the repetitive formatting and predictability of the encounter. If a couple knows that their lovemaking will take place on Saturday night at ten o'clock, perhaps after dinner and a glass of wine, and the intimate interaction always occurs as predicted, then it's not surprising that they begin to lose passion. (And if the encounter doesn't occur as expected, that can lead to problems too.) They can either resign themselves to a life of decreasing intimacy or begin a potentially damaging search for passion elsewhere. Seeking certainty with regard to honesty, fidelity, who's preparing dinner, and what time to pick up the kids from playdates makes life run smoothly, but subjecting matters of the heart and soul to the template of mechanistic certainty can lead to boredom and frustration.

Consider the word *uncertain*. Does it evoke a neutral, negative, or positive feeling? I believe that for most of us the answer is negative. "I feel uncertain about my future" should imply the possibility of a positive future as easily as a negative one, but ordinarily that statement would be seen as expressing fear or anxiety about what's to come. If I feel certain that my prospects will be disappointing, then uncertainty should provide a bit of relief. And yet, we don't use the word *uncertain* that way because we value certainty and are uncomfortable with uncertainty, except at a distance or as spectators.

Living life through the directive of certainty and predictability removes much of what it is to be human. After all, wonder and imagination don't comply with formulaic equations. This loss of enchantment contributes to any number of psychological and emotional challenges and in large part to the epidemic of anxiety and depression that our culture struggles with. Albert Einstein immersed himself in the uncertainty of wondering, his imagination piqued. He once wrote that at the age of sixteen he pictured himself chasing a beam of light, which ultimately led to his theory of special relativity. Such a breakthrough was possible only by not succumbing to the laws of a deterministic universe. When we are grounded in wonder and enchanted by the hopefulness of possibility in our lives, we are far less likely to feel depressed, anxious, or listless. The good news is that uncertainty provides the key to unlock the shackles of determinism.

CERTAINTY COMES UNDONE

The belief in determinism reigned uncontested from the seventeenth century until 1927, when the German quantum physicist Werner Heisenberg introduced his landmark uncertainty principle, also known as the indeterminacy principle.

By analyzing the existing data at that time, Heisenberg demonstrated that uncertainties, or imprecisions, manifested if one tried to measure the position and momentum of a particle at the same time. This revealed that certainty has its limits. He provided the example of using a microscope to locate a single electron, which would require bouncing light off the electron. The problem was that even a single photon of light would disturb the electron, changing its momentum and, consequently, its location. Both the location and the momentum could not be measured simultaneously. We could apprehend one feature only at the loss of knowing the other—showing that there appears to be a boundary to what we can know. The uncertainty principle transfigured science's "truth" about certainty.[1]

This turn in scientific thinking invalidated Newton's determinism, which held that all information could be accurately assessed and correlated. At the center of Heisenberg's discovery lies another deep implication: the loss of objectivity. The observer (the scientist and/or measuring device) intruded into the space of the observed (the electron), and as a result, the demarcation or separation between them fell away. This is also known as the observer effect. If we influence what we're looking at, we aren't truly standing apart and objectively observing.

The uncertainty principle leads to the notion of a genuinely interactive universe in which the observer affects the observed. If we are not separate from what we are looking at, we must be open to the consideration that our consciousness—informed by our beliefs, thoughts, and feelings—impacts and therefore alters what we are observing. The primary driver of reality is consciousness, not the material things of the mechanistic worldview. Rather than thinking of reality as "out there," separate from us, it would be advantageous to speak of "reality making." Reality is perpetually remaking itself, and we are an integral part of that process. We are not merely dependent on causal events that determine our present and future—as in the billiard ball analogy—or merely interacting with an independent reality, but we are thoroughly participating in the orchestration of that reality. The uncertainty principle shows us that our role in the reality-making process is paramount.

Over the last few decades, ample scientific studies have demonstrated that Heisenberg's uncertainty principle can be seen occurring in the macro world just as it does in the quantum realm. For example, in 2012, physicists at the University of Colorado demonstrated that Heisenberg's uncertainty principle prevails not only on the subatomic, microscopic level, but also at the level of visible matter. In a paper published the following year in the journal *Science*, the physicists describe how Heisenberg's uncertainty principle can indeed be demonstrated with objects large enough to be seen with the naked eye.[2]

In another experiment, when scientists observed beryllium atoms—which are indeed macro, not quantum—the atoms decayed more slowly the more the scientists observed and measured them.[3] Their decay was inexplicably altered by the act of observation. This nonrational, counterintuitive finding suggests that our very consciousness interacts with what we think remains distant and unaffected by us. We impact what we think of as "out there," and this realization provides us with deep meaning and purpose.

This is why embracing uncertainty is so valuable to our everyday existence. Uncertainty is the correlative of change and possibility. To live our lives to our potential and feel fully alive, we need to know that we play a role in what's to come. Certainty and determinism shutter out our participatory role in creating our future; uncertainty restores it. When we learn to invite uncertainty, we can facilitate desired changes in our lives by leaving possibilities open and accessible. Insisting on predetermined outcomes, we narrow our choices and often freeze in fear. The notion of predictability leaves us outside of the creative window as the formula for the future has been scripted. But when we accept that all of life is uncertain, from the microscopic to the macrocosmic level, we can recover our human potential.

THE FEAR OF MAKING THE WRONG CHOICE

I was working with a highly intelligent college student who had been excelling academically. But on a personal level, Antoine had a deep dread of making what he called "a serious mistake." He felt it would be a grievous error to choose the wrong major and career path. He was so immobilized by his fear of making a mistake that he dropped out of school altogether instead of risking a wrong decision. In his trepidation about the uncertainty of his future, he assumed that only one correct path existed. Antoine's acute analyzing of his circumstances caused him to lose touch with his inner voice, his intuitive wisdom. Choosing the "right" major is certainly an important decision, but it pales in comparison with dropping out of school.

Our fears about the future are often focused on what we call outcomes. But an outcome is nothing more than a momentary snapshot that we take in a particular moment. This way of looking at life—living in dread of certain outcomes—is rooted in the static picture of Newton's universe. From the perspective of a flowing participatory paradigm, no such thing as an outcome exists because reality is always continuing to unfold. If we are part of that flow, we can choose differently as we wish, but we must be in the flow. Embracing uncertainty frees us from fear of outcomes.

I helped Antoine look at his struggle through new eyes by sharing with him the new worldview of uncertainty—from which the fear of making mistakes is alleviated by the assurance of a multitude of possibilities. Once he could imagine a future that wouldn't have to be narrow but could widen as he grew, he released his fears and was readmitted to school.

To help Tom, the client mentioned earlier, surmount the anxiety caused by his overwhelming need for certainty, I explained the uncertainty principle and the scientific discoveries associated with it, and I shared its metaphorical implications for his fears about the future. He desperately wanted to change but couldn't as long as he saw the unknown as his enemy. I urged him to consider that the change he sought—to overcome his anxiety—could occur only if he, paradoxically, embraced uncertainty. Rather than resist uncertainty, he had to change his relationship with it by envisioning uncertainty as the ground of his genuine being, free from fear.

After a number of sessions, Tom took a one-month hiatus from our meetings. When he came back in, he reported—with almost an adolescent's enthusiasm—that he felt much better. I asked him to what he attributed this shift. He smiled and said his best friend was now uncertainty. It opened the road to his relief from anxiety. He then added the crowning touch—his new password for his accounts was "embrace," as in "embrace uncertainty." When we embrace what we fear, the fear dissipates. It loosens its grip on us.

WHO WILL I BE?

Angela had been mired in a loveless and emotionally abusive marriage of twelve years. Despite her pleading with her husband to enter marriage counseling, he continued to denigrate her and refused therapy. She felt depressed and hopeless, and although she often threatened him with divorce, she felt frozen. As she and I moved more deeply into our work together, it became apparent that her attachment to certainty kept her safe from the unknown. Even if that unknown—who would she be as a divorced woman?—was where relief might lie, a miserable certainty still felt safer to Angela. This was a clear example of our embedded bias against uncertainty.

I asked her to imagine standing with me on the bank of a river. I had her envision the current of the river as the flow of her life. I coaxed her to come into the river with me. Proceeding with the visualization, Angela walked into the river with me, but as we moved toward the middle—where the current became stronger—she reported that she saw herself grabbing onto a large rock. I asked her to let go and embrace the current, to come along for the ride. Balking at this, she looked ahead and saw a bend in the river, protesting, "But I can't see where the river will take me. I need to know."

I told her we're not supposed to see around the bend. I explained that she was not powerless but could navigate the currents (her future), making adjustments as required, trusting that her life would evolve, as it should, with her at the helm. But it was essential that she enter the current, the metaphorical flow of her life. She needed to accept the fear and uncertainty of what lay ahead of her as a divorced woman rather than the daily dread that she experienced in her marriage.

Like many people who anticipate making significant changes, Angela experienced fear when contemplating the question "Who would I be?" She would certainly be experiencing her life differently—and this provoked further fear. Under her current circumstances, Angela could fault her husband for her unhappiness and continue to feel like a victim. Clinging to an unhappy yet certain state is essential to playing the role of victim. The certainty of the present locked her in despair but paradoxically kept her from becoming responsible for her own life. However nonsensical it may sound, we often cling to the dysfunction and unhappiness of the certain present to avoid the uncertainty of the future. To use the cliché, we choose the devil we know.

As I helped Angela to envision the ways she would like to experience her life—with neither the weight of a disastrous marriage *nor* the fear of the unknown—she became excited by the possibilities. It took her awhile, but she finally freed herself from both her fear of uncertainty and her destructive marriage.

When you struggle with unsatisfactory or troublesome aspects of your life, ask yourself, "How is my fear around the unknown and its uncertainty getting in the way of my change process?" Imagine yourself welcoming and embracing the uncertainty, and you'll gain a sense of self-empowerment that can free you from the grip that certainty may have on you.

RIDING THE WAVE OF CHANGE

The commonly held belief that it's hard to change has its roots in the deterministic view that we are inert and static and that moving past our stuck state requires an uphill effort. From the old worldview of certainty, Newton's first law of motion might be stated, "An object at rest stays at rest, and an object in motion stays in motion at the same speed and going in the same direction unless acted upon by an unbalanced force." When we consider ourselves to be the object, we conclude that we can't move, stop, or change direction unless acted upon by an outside force—that it is hard for us to change. This conclusion makes perfect sense and becomes a self-fulfilling prophecy.

Because our beliefs around change ordinarily instruct us that it's either difficult to achieve or unwanted, we see change as problematic at best—an anomaly, the exception not the norm. While change *is* challenging and the exception within the mandate of determinism, it's easily accessible and normal in the quantum worldview. According to quantum physics, all of reality appears to be perpetually recreating itself, to be engaged in a dance of inexorable movement. Nothing and no one is fixed and inert; everything and everyone is part of a never-ceasing dance of reality making.

Uncertainty frees us from the harness of determinism and lets us join in the flow of possibilities, free from the limitations of predictability. Possibility invites and requires our participation. To envision and actualize the future we long for, we must view uncertainty as our ally.

3

RECOVERING OUR
LOST POTENTIAL

Our sense of self is constructed early in life, sometimes through traumatic events and at other times more subtly. An aspect of quantum physics called *wave collapse* can illuminate how this construction occurs and, more importantly, how we can empower ourselves to live a life that is unburdened by our past.

One day in my office, a client named Jill recalled the words her mother spoke to her when she was about eight years old: "When I learned I was pregnant with you, I told your father I didn't want another baby." Despite the fact that her mother was otherwise devoted and loving, Jill's acutely personal takeaway was damning. She felt unwanted and therefore unlovable, then and ever since. She carried this core belief with her throughout her life. Her own inner monologue was perpetually self-critical, confirming her belief that she wasn't lovable. The snapshot Jill had taken of herself early in her life had become etched into her psyche as her embedded truth.

Jill's belief affected her relations with her husband, children, and friends. Notwithstanding her husband Bob's loving devotion to her, Jill questioned his loyalty and truthfulness in light of seeing herself as unlovable. Her belief about herself was becoming a self-fulfilling prophecy: she was forcing Bob to withdraw his love as his frustration mounted. What Jill experienced is not uncommon, for what we believe to be true about ourselves—and others—contributes to our reality-making process. Prior to her mother's remark, Jill's identity could have evolved in limitless ways, but that range of possibility became narrowed by that one short sentence.

For virtually all of us whose beliefs have been ingrained with the mechanistic worldview, the world as seen through quantum physics appears to be suffused with a kind of nonrational strangeness. One of the fundamental aspects of the quantum worldview, for instance, is that elementary particles exhibit a somewhat "schizophrenic" nature. I use that word not in its complex clinical sense but in the conventional meaning of having a "split personality"—and every quantum entity indeed has the dual capacity to exist as either a wave or a particle. Physicists refer to this tendency as the *wave-particle duality*—a notion that rubs against our commonsense logic. Ordinarily, we believe that things either *are* or *are not*, that they are distinct in their nature. This either-or thinking can also be referred to as *binary thinking*, which leaves only two distinct paths open to us. Binary thinking is a major aspect of how we observe and construct reality. Yet this either-or reality apparently doesn't apply in the quantum realm and is questionable in our everyday lives as well.

The quantum reality exists in what is known as a series of probability waves, with an infinite number of potential outcomes. This means that when the particle is not being observed, it exists as a *waveform*, which in quantum language represents a state of pure potential, known as *superposition*. This term proposes that as long as we do not know what the state of any object is, it actually exists in all possible states simultaneously, as long as we don't look to check. In that sense, the wave represents pure possibility. The very act of observation reduces the wave (potential) to a fixed thing—a particle. This reduction is referred to as *wave collapse*.

All that may sound far removed from our day-to-day world of personal relationships, fears and anxieties, love and hate. Yet a similar thing occurs in our lives. When we have particular experiences and make certain observations of ourselves, or have them made of us—typically in childhood—we experience the psychological equivalent of a quantum wave collapse.

As newborns or infants, if not at conception and in utero, we resemble the infinite possibilities of the wave; our personality, not yet defined, is in a state of potential. Notwithstanding matters of genetics, environmental influences, or considerations of archetypal, astrological, or karmic influences (however we may feel about those concepts), our identity is not yet determined and fixed. But before long, we move from the potential of the wave to the "thingness" of the particle. The personal evolution of our personality gets stunted, and our growth becomes fitful. How does this happen?

Ordinarily, even a single yet significant experience is sufficient to collapse our personal wave of potential. Jill experienced a powerful wave collapse after her mother spoke one particular sentence to her. Sometimes all it takes is a hurtful statement or an embarrassing experience in our early years to reduce the potential of our personality to a narrow, restricted self-image. These events need not be traumatic; they may, in fact, be subtle. Yet in those moments, our potential fades. It's as if we have taken a snapshot of ourselves, and we become frozen in time. I refer to these as *confining* wave collapses, in contrast to the *defining* wave collapses that usher in defining moments. We are no longer the potential of the wave but the finiteness of the particle. And we carry this picture of ourselves with us through our lives, allowing it to burden and limit us. We lose the authorship of our life story.

The initial wave collapse sets up a recurring incidence of similar experiences as our beliefs about ourselves and others become self-reinforcing. What we think of ourselves shapes our interactions with both others and ourselves. This habit obstructs our ability to change or evolve as we cling to our perceived "truth" of who we are. The themes of subsequent collapses may vary, but they are often self-limiting if not denigrating. We generate thoughts such as "I'm not good enough" or "I'm not smart enough" or even more simply, "I'm not loveable." The actors who perhaps unwittingly participate in scripting our personal beliefs are often our parents, but they may also be teachers, friends, relatives, or even strangers. We cling to these habituated beliefs about ourselves in accordance with our primary wave collapses. In spite of new events that should cause us to reconsider or reevaluate our beliefs—for example, Bob's devotion to and love for Jill—we remain rooted in the way we see ourselves. We become embedded in the groove of our self-referencing beliefs and block the opportunity for growth and change.

Less dramatically than the hurtful comments or abusive actions of others, the patterns of family dynamics may cause us to acquire certain personality traits. These influences tend to be more chronic and may fly beneath our radar screen. If you grew up in a highly conflicted or alcoholic family, for instance, you may have coped by developing a "people pleaser" or peacemaker persona. We can think of this chronic condition as an extended wave collapse rather than the result of the acute single event.

I had referred another client, Helen, to meet a colleague of mine so that they might explore matters of mutual professional interest. Helen made an appointment to meet with Jim at a conference he would be attending. When Helen arrived, Jim was engaged in conversation with others and didn't notice her waiting to introduce herself. Shortly thereafter, Jim left the conference without acknowledging Helen.

"I guess I just wasn't important enough for him to wait to meet me," Helen said to me later. She reported this as though it were an established fact instead of an interpretive opinion of her own construction. I asked her how she knew this to be true and whether other explanations might apply. For example, I knew Jim to be notoriously absentminded; he might simply have forgotten he was supposed to meet her at the conference, or he may not have noticed her waiting to introduce herself. In fact, he had overlooked appointments with me in the past, and I certainly didn't conclude that he viewed me as unimportant.

I suggested that Helen saw herself as not valuable, and she was projecting that insult onto Jim, thinking he saw her as she saw herself. At first she resisted this possibility. She related some compelling stories, which revealed that as a child she had felt like her mother's servant and that her entire childhood was about her dutiful obedience to her mother. Helen waited on her mother hand and foot as their parent-child roles were reversed. She was deprived of the value that every child deserves, and so she basically felt unimportant. This wave collapse had disastrous effects on her self-esteem. Throughout her life, her thoughts almost *automatically* continued to affirm this affliction.

Our primary beliefs about ourselves that have been generated by our wave collapses orchestrate the quality and nature of our thoughts, which make specific the general theme of the wave collapse. If, like Helen, our core belief is that we are not of value, we can predict the kinds of thoughts we might then experience, such as "I'm not important" or "Why should he pay attention to me? I don't matter." These thoughts measure ourselves against others, and the predictable result is that we see ourselves as subordinate to others. These thoughts and their resulting feelings can leave us trapped inside a self-induced container of low self-esteem.

RETURNING TO PURE POTENTIAL

To free yourself from repeating harmful and confining wave collapses, consider this Possibility Principle: In the nanosecond before your next thought, you are in a state of pure potential.

In the space between our thoughts, we are similar to the wave—full of possibility. Once we attach to our next thought, the ensuing wave collapses, and we create our reality in that moment. If we continue to have self-limiting or injurious thoughts, we remain adhered to the damaging effects of the primary wave collapse. In therapy, a client often experiences a breakthrough, a significant moment during which a highly anticipated insight becomes illuminated. This event presents a new state of potential and, with it, the possibility of a defining moment, in which the client can break into new terrain. The person will then select which reality to summon by thinking either "What a relief! I've broken through" or "What's wrong with me? Why has this taken so long?" One thought is self-affirming and offers relief and the possibility of vaulting forward, while the other is self-critical and resists progress. The thought you select will chart your path.

Simply put, the thought we engage will summon the reality of our next moment. We can move forward in breaking new ground, or we can summon an old familiar thought, abandoning the insight.

Obviously, we can choose vastly differing experiences. The potential is all that exists prior to the next thought. Our struggle with change is in part caused by our habituation to old thoughts.

Helen's confining wave collapse of feeling devalued set in motion her lifelong inner narrative. I could see the automatic nature of her thought coming through in the statement, "I guess I'm not important enough for him to wait for me." This type of thought, instinctive and programmed, burdened and afflicted Helen. I worked with her to identify that thought and separate from it so that she could see what it was telling her. This established another essential principle for her: if you can learn to see the thought, you don't have to become the thought.

As we talked further, I asked Helen whether she had attempted to contact Jim after the conference. She continued to cling to her story that there wasn't any point in doing so because she had been blown off. I suggested that her thought had indeed *told* her that she was blown off, but how could she actually know that for a fact? What she was doing was creating a storyline to conform to her beliefs and her personal history. Was she perhaps simply replicating wave collapses from earlier in her life? Was this her thought stuck in an old groove? When pressed, she couldn't offer proof that her story was true. If we can't know for sure that our story is true, then we need to look at how we confuse our story with the truth of a given situation. I asked Helen if it was plausible that she had believed an untruth about herself her entire life. She reconsidered and acknowledged that she might be personalizing the events around her meeting with Jim to conform to her self-image.

When I explained to Helen the theory behind wave collapse, she quickly grasped the concept. I had her envision an alternative and positive wave collapse in which her mother had been maternally inclined and actually doted on her. What then might her belief about herself be? She permitted herself this alternative point of view and considered that perhaps she wasn't irrelevant. Doing so meant that she also needed to embrace her discomfort as she moved beyond the limits of her familiar zone. (As much as we may suffer with our limiting beliefs, we often feel dissonance between our old beliefs and new ones and apprehension about stepping into the new potential of who we may become when we release the old beliefs.) Our work then focused on helping Helen break free from her addictive tendency to malign herself—or more precisely, the tendency of her *thought*.

Helen's progress from that point was impressive. She began to value herself, and this shift in her self-esteem led to her experiencing her relationships in an entirely new way. By severing the grip of her old identity, she gave herself the opportunity to select a new and positive wave collapse with self-affirming thoughts. Breaking free from the predictability of the confining wave collapse requires uprooting the repetitive thoughts that inform the old belief. Noticing the repetition of these old thoughts, in lockstep with the wave collapse, enables the shift.

The quantum view of the universe tells us that reality appears to unfold perpetually from a state of potential—what we earlier called *superposition*. To access the universal potential, we must devote ourselves to apprehending that possibility, which lies in the instant prior to collapsing the wave with our next thought or feeling. Our thought literally summons our construed reality. Thoughts that emanate from the habitual groove of old wave collapses are likely to re-create more of the same feelings and experiences. Our thoughts therefore *re*-present our past experience. This is why we struggle with change. Taking a new snapshot and actualizing new thinking will, however, script a new experience, allowing us to participate fully in an evolving reality.

RELEASING YOUR PAST

Appreciating how the wave collapses in our lives informed our sense of self is essential in priming the pump for change. Often the meanings we attribute to the events of our personal history prevent us from creating effective change as we are reduced to seeing ourselves as victims. Many adults have memories of an abusive, loveless, or disappointing childhood because they didn't receive the nurturing and love that is every child's birthright. But if we choose to keep focusing on limiting events of our past, then we choose a present that predicts a similar future. At some point, we need to stop choosing to believe the meaning we ascribed to our past and script a different present.

I am not suggesting that we either avoid or suppress painful memories. By all means, we need to bring them into the light and process them, so we can loosen their grip on us. The goal, though, is to disarm them and eventually release them. The thought that we choose in the present moment is almost entirely responsible for who we are in that moment. If we continue to summon the same habitual thoughts, we won't realize the potential that awaits us.

What we are seeking are new wave collapses that implant *positive* self-reflections and identities as we grow past the grip of the negative ones. These new defining wave collapses offer the route for our defining moments.

In working with Jill, I asked her to consider what she could have said when her mother told her that she hadn't wanted another baby when she found out she was pregnant with Jill. Jill responded, "That makes me feel terrible and unwanted." Had she actually said that at the time, she might have experienced a positive defining wave collapse. Simply reliving that encounter in a way that gave her some power was part of her healing process. By expressing her feelings, she was able to establish a worthier sense of self. Instead of thinking, "I am unlovable," which speaks to a fixed state of being, she could reframe her belief to "I've thought of myself as unlovable, and now I know why." This belief is amenable to change.

Think of a core belief you hold about yourself that greatly limits how you experience your life. We think of these beliefs as our "truth." They may range from "I'm not smart enough" to "People just don't respect me." Or they may sound like "I'm a poor communicator" or "Conflict makes me uncomfortable, so I avoid confrontation."

After you've identified that core belief, ask yourself how you came to this belief. You might recall an embarrassing, shameful, or traumatic moment from earlier in life in which this "truth" took hold. For example, you raised your hand in class to ask a question, and everyone laughed at how silly your question seemed. So you decided to never risk that kind of exposure again. Now you play it safe and really think long and hard before you speak. Or your belief could have been caused by more chronic circumstances, like growing up in a volatile home and coming to believe that you couldn't really share how you felt out of fear of catalyzing violence.

This is the confining wave collapse that has created your limiting belief about yourself. If it was caused by a specific moment in which you felt shamed or ridiculed, picture yourself back in that moment. Then imagine yourself finding your voice. Tell those involved how you feel about what just happened or what they said or did to you. Finding your voice in this way helps release you from the bondage of the confining wave collapse. Reflect on how different your beliefs about yourself would be if this event had never happened.

If your beliefs about yourself were informed by chronic rather than acute circumstances, such as having an alcoholic or abusive parent or growing up in a volatile home, remind yourself that these beliefs are a product of your experience. Imagine yourself being raised in a loving and supportive family. Now how differently might you feel about yourself? Once you choose to reclaim your potential, you cease being a victim of damaging circumstances. Say to yourself, "I don't need to be imprisoned by my past if I choose to free myself from my limiting beliefs."

You are more than your experiences, and an infinite potential awaits you as you allow your identity to evolve. Once you learn to see how your beliefs are informing you, you are free to break into new terrain and achieve a defining moment. Witness your thoughts and recognize the story they are telling you. Don't confuse them with the truth. You can learn to rewrite your story.

Many individuals were fortunate enough to have experienced exemplary defining wave collapses that affirmed them and helped them to secure a strong sense of self. This typically leads to healthy self-esteem that enables them to craft their personalities and experiences free from constrictive encumbrances. But those of us who haven't yet experienced such a gift can *learn* how to overcome our burdens and reach the full range of our possibilities. The way we've been trained to think that change is hard or implausible once again grows out of our operating worldview. This belief in inertia developed from the causal determinism of the mechanistic worldview in which our present and future are dependent on our past. The new principle derived from the participatory worldview—potentiality—invites us to free ourselves from aspects of our past that don't serve us. We needn't stay stuck in the fixed state of the particle but can ride the possibilities of the wave.

HOW WE ARE ALL CONNECTED IN AN INSEPARABLE, PARTICIPATORY UNIVERSE

W hat may be the most provocative discovery of modern science remains relatively obscure to the general public. Perhaps this is because it has been too radical for us to embrace. It shatters our picture of reality—and subsequently how we see ourselves operating in the world. If we did open fully to the implications of this finding, it would compel us to reframe our entire way of thinking. Yet, paradoxically, we would also be much more likely to flourish because this discovery helps us grasp just how thoroughly interconnected we all are—with each other and with the universe. We would have reason to embrace the age-old spiritual principle that "we are all one," but based on new scientific reality.

In 1935, Albert Einstein and the renowned Danish physicist Niels Bohr engaged in a theoretical debate that would ultimately shape our understanding of what reality looks like. Einstein proposed a thought experiment—known as the Einstein-Podolsky-Rosen (EPR) paradox—and it became a hotly contested theoretical battleground between the two intellectual titans.[1] The thought experiment related to the behavior of a pair of photons, which are simply light waves that have become tiny particles. When the two particles are created at the same point and instant in space, they become entangled as a pair and experience what is known as a shared state. Paired photons have opposing spins or rotations. If particle A, for example, spins in a clockwise rotation, the spin of particle B must be counterclockwise. The central question of Einstein and Bohr's debate revolved around what would happen if an enormous distance separated the particles—imagine half a universe—and the spin of particle A were altered to counterclockwise. Both physicists agreed that particle B would necessarily change its spin accordingly. But how long would that take to occur?

Einstein argued that the time required for one photon to signal the other could be calculated based on the distance between them and the laws determining the speed of light. Bohr, on the other hand, boldly predicted that no signal would be necessary from one photon to the other, and hence, no time would elapse before their spins respectively reversed. He claimed that because both photons continued to exist in an entangled state, they were still inseparable, regardless of how distant they were from each other. In scientific parlance, this entanglement regardless of distance is known as *nonlocality*.

Bohr's claim flew in the face of classical Newtonian physics, which mandates that causality prevails and time must elapse for the signal to be communicated: orthodox cause-and-effect determinism. Bohr was proposing that, in certain circumstances, an entanglement exists with no separation, notwithstanding the vast distance between the photons. More to the point, Bohr was proposing that the universe is fundamentally an inseparable, undivided whole.

Einstein vehemently objected to this seemingly nonrational claim. Despite the revolution he catalyzed with his theory of relativity, Einstein still maintained some of the tenets of Newton's classical paradigm. He insisted on a discrete and objective realism that required separation. His point of view posited what was called a *local reality*—separation—subject to classical time and measurement. He protested Bohr's notion of an acausal reality, remarking, "If I should be born again, I will become a cobbler and do my thinking in peace"—his way of saying that he would abandon his belief in science if there weren't a classical separation of space and time, with cause and effect.[2] He referred to the phenomenon of entangled photons as "spooky action at a distance."[3]

The debate raged on for decades and divided physicists into opposing theoretical camps. In the early 1960s, Irish physicist John Stewart Bell developed what became known as Bell's theorem, a formula that seemed to not only resolve the argument but also prove Einstein correct.[4] Nearly twenty years later, the technology was finally available to test Bell's theorem—and Einstein was proven wrong! No signal was required to travel between the photons to alter their spins. Communication was instantaneous, and the possibility of hidden variables that could account for this—Einstein's suggestion—was disproven.

The photons were entangled as though they were still as one, just as Bohr had postulated. This hypothesis has been retested countless times, always with the same result. A 2015 article in the *New York Times* reports that "since the 1970s, a series of precise experiments by physicists are increasingly erasing doubt—alternative explanations that are referred to as loopholes—that two previously entangled particles, even if separated by the width of the universe, could instantly interact."[5] As recently as 2015, an experiment in the Netherlands further confirmed inseparability.[6]

So as radical as it may seem, under certain conditions the universe appears as an undivided, inseparable whole. The philosophical inference drawn from this conclusion suggests that determining what is independently objective and real—Einstein's claim—seems somewhat murky, if at all achievable. Not only does this discovery of an inseparable universe appear to invalidate the bedrock principle of classical physics known as *locality* (which says that an object can be directly influenced only by its immediate surroundings), but it also opens the horizon to a spectacular contemplation of unbroken wholeness. The cornerstone of Newton's theme of separation crumbles and is replaced by inseparability.

Increasing evidence indicates that entanglement occurs not just between photons, but in the larger macro realm as well—undoubtedly affecting humans. The June 2011 cover article in *Scientific American*, titled "Living in a Quantum World," proposed that larger biological entities were amenable to entanglement, which had been witnessed in living organisms.[7] A number of studies tell us that the macro world is, in fact, thoroughly quantum, and many scientists have echoed this. Ervin Laszlo, in his book *The Connectivity Hypothesis*, refers to this entanglement as a state of coherence and suggests that not only the quantum but also the macro and cosmic realms all operate from this inseparability: "The kind of coherence observed in the domain of the quantum was believed to be limited to that domain; the world of macroscopic objects was thought to be 'classical.' Yet this assumption is no longer entirely true. There is growing evidence that an anomalous form of coherence also occurs at macroscopic scales; indeed, even at cosmic scales."[8] The division between the quantum world and the macro world seems to be an artificial distinction most likely created by our minds.

The phenomenon of entanglement may well explain the occurrences of distance healing, remote viewing, psi phenomena, the power of prayer, and telepathy. Let's consider another pair of twins, but this time humans instead of photons. She lives in New York City, and he lives in Paris. One day, as she is getting out of bed, she walks toward the shower, slips, and breaks her ankle. Precisely at that moment, her twin brother in Paris feels an excruciating pain in the same location in his foot. Just as with the photons, there is no signal sent from one to the other. They are each—at least momentarily—part of the same whole, so to speak. They are as entangled as the photons. We tend to regard such an anecdote as inexplicable, just one of those odd things. Those who cling to classical science may try to justify this phenomenon by some explanation of shared DNA. Yet these occurrences also happen among people who aren't twins and may not even be related. Have you ever thought of an old friend with whom you've lost touch and to whom you haven't spoken in years—and within a few minutes, or in that very instant, you get an email or a phone call from that person? Simple probability doesn't explain such events. Something much deeper is occurring: inseparability.

A number of scientific studies have revealed these events to be more common than we may tend to believe. Throughout his book *The Sense of Being Stared At*, British biologist and researcher Rupert Sheldrake draws on more than 5,000 case histories of "apparently unexplained perceptiveness by people and by nonhuman animals," including the kinds of "telephone telepathy" I referred to above.[9] Such events aren't unusual; they merely lie outside of our typical field of vision. They *seem* to appear rarely, rather than routinely, only because we don't recognize them; our way of thinking tricks us into seeing separation where it doesn't exist. The mechanistic filter through which we see obscures connectivity and deludes us into seeing separation.

When we marginalize these and other examples of wholeness as simply being inexplicable or weird (or to borrow Einstein's word, *spooky*), we disconnect from the transcendent experience and, in doing so, do ourselves a great disservice. When something occurs that doesn't fit our operating belief system—derived from the mechanistic worldview—we must reexamine our beliefs, not discard the experience. Good science requires looking at these anomalies, not deflecting them. When we embrace the dissonance and ensuing confusion around events we can't readily explain, old paradigms fall away and new worldviews emerge. In this circumstance, science broke new ground many decades ago, yet we haven't integrated the science and its larger implications into our everyday lives.

The implications of entanglement necessitate a radical reconsideration of the way we envision reality, but just as importantly they necessitate that we overhaul the way we think and envision ourselves. Such a mind-altering reality defies our commonsense approach to cause and effect, which naturally require separation, objectivity, and causality.

MISSING THE BIG PICTURE

Once again, the way we think reality operates orchestrates how we operate. As Polish philosopher Henryk Skolimowski suggests, "Our lives are the mirrors in which the fundamental characteristics of the universe, as we understand it, are reflected."[10] Our rootedness in Newton's mechanistic paradigm compels us into separation. The resulting beliefs create a sense of isolation and an ensuing loss of meaning, which explains why depression, greed, inhumanity, failed relationships, and ineffective communication are so prevalent in our world today.

Our mechanistic worldview also fragments our thinking, which ruptures our connection with the inexorable flow of the universe. René Descartes asserted that the most efficacious way of attaining knowledge is to break up the whole into smaller parts, each part being more analyzable than the whole.[11] This philosophy led to the notion of divisibility by thought. Descartes proposed that any big picture should be sliced and diced into smaller pieces, and once these smaller components were analyzed, the whole big picture could be reconstituted.

The problem with Descartes's conception is that the whole is not merely the sum of the parts. The whole expresses a tapestry that is irreducible—a vital energy force unto itself. If we examine any of the billions of parts of the human body, for example, we could never reassemble individual humans or capture their uniqueness. Nor could we disassemble and then reassemble a rainbow, an insight, or the experience of love. Machines can be broken down, analyzed, and understood, whereas more complex organisms, including humans, cannot. When we seek to see the whole, we may at our discretion choose to analyze a part or parts of it for some efficacious purpose and then return to holism; but when our minds are trained to analyze parts as if on autopilot, we lose the capacity to perceive wholeness.

Our mind constructs divisions and then denies that it has, in fact, done just that. Thought dissects and splits asunder the natural order of inseparability. It is as if we go through life with binoculars affixed to our eyes, witnessing only the narrow visage in front of us and detached from the grander view. From this fragmented view, we unwittingly create enormous disharmony in our lives. As a result, we become victims to circumstances that we believe external to us as we obfuscate our own involvement.

The Nobel Prize–winning physicist David Bohm explained fragmented thought as the filter through which we see differences and divisions, as opposed to a more organic way of organizing our thinking. As I've noted, focusing on specific parts of the whole for utilitarian purposes has its benefits, but only in certain contexts. Analyzing a human cell under a microscope may well elicit valuable information. Even then, however, the data must be looked at in the context that the cell operates as part of a system of billions of other cells, each influencing the whole.

In the medical field, we can easily see the fragmented way in which we break down and separate various disciplines and fields of specialization from one another. This creates what is known as "silo thinking." Each medical specialization may operate blindly in regard to other disciplines, without sharing critical information. When this occurs, the specialist treating a particular ailment may indeed be interfering with, if not exacerbating, a different medical challenge. The whole person and their wellness might be obscured through this fragmented approach. It's not atypical for a specialist to prescribe medication—let's say to lower your cholesterol levels—and neglect to offer that the meds may have deleterious effects on your memory or your heart. So we exchange one problem for the other. In her provocative book *The Watchman's Rattle: A Radical New Theory of Collapse*, the sociobiologist Rebecca Costa examines the perils of such thinking; she describes the ways in which a symptom-obsessed society is constantly looking for a quick fix and easy cure when it should be doing the hard work of seeking out deeper, longer-lasting solutions.[12]

By contrast, an approach that operates from wholeness would address the entirety of the person, system, or problem. During his 2016 State of the Union Address, President Obama called on Vice President Joe Biden to lead a national initiative to eliminate cancer as we know it—referred to as the "cancer moonshot."[13] To achieve this goal, the president proposed that the fragmented and separate silos of scientific research coalesce in a unified collaborative effort to vanquish cancer. This approach is a step forward, but for it to have real efficacy it would have to take into consideration all the ways in which we have created the very disease we seek to conquer. Recognizing how we had a hand in creating the problems that we see as "out there" is central to defragmented thinking.

My client Roger suffered from anxiety, obsessive thinking, and depression. Not surprisingly, his relationships and career were also suffering. He was miserable. At the root of these afflictions was, I believed, a single fundamental problem: low self-esteem. Before I could encourage Roger to look at this issue, I had to work to free him from his addiction to analyzing, to looking at the parts of his problems rather than the whole (his low self-worth).

He was constantly searching online—on WebMD and other such sites—for the *cause* of his malaise. Was it due to a food allergy or traces of chemicals that might be harmful to him? His research even led him to explore whether his depression might be caused by an allergy to his mattress. I challenged him to release his belief that his cure lay in one of these areas and to take up a larger issue: that he didn't value himself. He had to look at the whole, the big picture, and so we focused on the way his thoughts split off and overanalyzed, as he obscured the real issue of how he felt about himself.

Roger began to appreciate what I was proposing, and we eventually settled in to treating his core challenge of low self-esteem. My message was not to focus on what he thought so much as *how* he thought. He might have food sensitivities or allergies, but the first step was to stop looking "out there" for the cause. Much of our emotional and mental distress is the by-product of fragmented thoughts, as we disconnect from others, our own self, and the universe at large. The result is an existential despair that develops as we experience life cut off from the inseparable universe.

On another level, our sense of spiritual grounding is ruptured by fragmented thoughts. The absence of a more congruent sense of meaning and purpose leaves us spiritually bereft, once again as individual, separate parts in the machine. Because the universe appears to be a flowing whole, in which all parts interpenetrate and inform one another, our thinking needs to coalesce, defragment, and become coherent as well. Such coherence is the wellspring of health and resilience.

Imagine yourself as a molecule of water in the ocean. You see your neighboring molecule as separate and distinct from you. Given your lack of perspective, you miss the larger picture: both you and the neighboring molecule are part of the same wave. Yet while water molecules may appear separate through the eye of the microscope, they actually interpenetrate each other as part of an indivisible whole. Each molecule is an integral part of the wave, and each wave is part of the ocean.

The same truth applies to humans. Each of us is extraordinary in our uniqueness yet a constituent of the seamless, inseparable whole—not only the whole of humanity but of the entire universe as well. If no separation exists among constituent parts of the universe, then all things and beings interpenetrate the whole. Everything informs everything else, which includes humans, an intrinsic part of the galactic tapestry.

When we learn to see and participate in that wholeness, remarkable things happen. I caught a glimpse of the possibilities while viewing CNN during the millennium celebration. For twenty-four hours, every country on the planet celebrated the dawn of a new millennium. As I watched, it occurred to me that for this span of one day, there were no divisions between countries—only one planet, turning in its rotation toward the birth of a new age. I also contemplated that countries aren't intrinsically real. We did, after all, make them up. When we view Earth from space, we don't see a geopolitical map; we see something closer to the natural topographical layout that we all inhabit. Our current political system of nations is only a product of thought rooted in separation and played out through wars and treaties.

Recognizing that we are an integral part of that new vista of reality offers us innumerable advantages. Meaninglessness is replaced by a bounty of meaning and purpose as we regain our sense of connectivity. Little could be more monumental than our shift of mind.

WE PARTICIPATE IN THE CREATION OF THE WHOLE

This emerging paradigm of inseparability suggests that the universe is thoroughly participatory. If, as quantum mechanics describes, the universe operates from a flow of unbroken wholeness, then the fundamental reality is participation. To be an indivisible part of the whole is to participate in the whole.

This participatory worldview leads us to the process of *becoming* as opposed to the condition of *being*; we and all matter are involved in a dance of movement rather than existing as fixed states. The Greek philosopher Heraclitus said it perfectly: "Everything flows and nothing abides."[14]

Contrast this notion of participation and perpetual flow with Newton's fixed, machinelike reality, in which everything and everyone is separate and inert unless acted upon by some force. Flow invites personal growth and evolution so that the change process, far from being difficult, is immediately accessible with a shift of your mind. The absence of human significance in the classical worldview leaves humans here quite by accident, as part of the machinery. The participatory worldview, informed by inseparability, places human will and consciousness in the driver's seat of life.

Since flow suggests perpetual movement, "truths" and "laws," which reign supreme in the classical worldview, lose relevance as subjective experience replaces the myth of objectivity. The classical mechanistic approach demands that we quantify and objectify; the participatory ethos refers us to movement, process, and subjective experience.

To clarify, separation is synonymous with *objectivity*, in which we can stand apart and observe what is "out there." By contrast, inseparability reflects a participatory *subjective* view of the universe. John Archibald Wheeler, the celebrated theoretical physicist, wrote, "The universe does not exist 'out there' independent of us. We are inescapably involved in bringing about that which appears to be happening. We are not only observers. We are participators. In some strange sense, this is a participatory universe."[15]

Wheeler's description captures just how significant and powerful we are. In our participatory role, we orchestrate much of what we experience. With every thought, feeling, and action, or lack thereof, we impact the whole. Just as throwing a rock into a pond affects the entirety of the pond, creating ripples that interact with each other, so each of us creates ripples of consciousness that affect all the other ripples and the whole of reality. Our thoughts, feelings, and actions—or lack of them—are essential to the reality-making process. Viewed this way, our lives become profoundly meaningful and purposeful. We matter.

SEEING THROUGH WHOLENESS INDUCES EMPATHY

When we see reality through this new prism of indivisible wholeness and participation, we come to understand that doing harm to another is to do harm to our own self. Committing violence to another should be as ludicrous as your left arm attacking your right arm because it thinks the right arm is separate and in opposition to itself, not part of the same body. In the worldview of inseparability, the distinction between other and self melts away, allowing compassion and empathy to surface as primary emotions.

Whereas *empathy* usually suggests the ability to "get into another's shoes" and to imagine how they feel, inseparability takes this concept several steps further. If the other and I are part of the same whole, we have much more in common than I might have imagined. Part of my being is entangled in them as they are in me; this realization induces empathy to flow effortlessly. This type of empathy is precisely the foundation for thriving relationships, which we'll look at in detail in chapters 10 and 11. We are dearly missing empathy in our lives as we rush past one another in our separation-driven self-focus. Empathy gives oxygen to our relationships and sustenance to our lives. It is the thread that connects us and is the source of our humanity. Imagine how wondrously life would change if our emphasis shifted from "me" to "we." Given inseparability, the golden rule is no longer simply a homily but a logical extension of oneness.

In the classical orientation of separation, our sense of individuality reigns supreme, which naturally engenders the competitive drive. I had been working with Roberto for some time, primarily because of significant complaints that his wife and children presented. They found him to be controlling, disconnected, and generally miserable. Roberto lived in the extreme of individualism. His drive was to assert his needs; he appeared to be what we'd call selfish or self-absorbed, and so compassion lay pretty much beyond his reach. Empathy was a construct that made little sense to him, for he viewed it as self-sacrificing. In his drive to satisfy his own insular needs, Roberto was fully disconnected from his family. His priorities were work and achieving ever-greater financial success. At various times he defended himself by saying, "It's a dog-eat-dog world" or "It's every man for himself." Insignificant discussions about what movie to see or where to go out for dinner broke down into arguments as he competed to assert his choice and "win." How could he connect, collaborate, or truly love when he felt so imprisoned by his sense of separation and competition?

Taken to an extreme, the compulsion to compete, to win at all costs, is pathological and contrary to humanistic values. You can't seek to win and to relate at the same time. You can't experience compassion, understanding, or empathy if your goal is to outpace others and cross imaginary finish lines before them. As Roberto had discovered, in personal relationships the need to be right often precludes harmony and understanding. This philosophy propels our lives in an endless race and destroys our ability to be present and nurture others and ourselves. And it does unthinkable harm to our children. A 2016 *New York Times* article entitled "Is the Drive for Success Making Our Children Sick?" reports alarming rates of depression and anxiety in high school students, which are "a microcosm of a nationwide epidemic of school-related stress."[16]

I shared with Roberto my belief that he was struggling with his relationships in large part because of his worldview. Given that he was rooted in his intellect and interested in science, this discussion interested him. Since it wasn't personal, he didn't need to defend himself when I proposed it, and he eventually opened to rethinking his prevailing attitude about individualism and competition, to the benefit of his family and himself.

Imagine yourself interconnected with everyone and everything—just for one day to start. See a part of yourself in them and a part of them in you.

Start by visualizing the people you know best, your family and loved ones. Then expand to include your colleagues and people you encounter casually at the store or bank. Finally, extend your embrace to include people you consider enemies and the countless people around the world you will never meet.

THE VALUE OF COOPERATION:
PLAYING OUR UNIQUE ROLES IN THE WHOLE

When we open to inseparability, the cooperative spirit ultimately subsumes the competitive divide. This idea can seem threatening to some people because a commonly held belief is that cooperative energy diminishes achievement, innovation, productivity, or all three. This metanarrative is at the heart of industrialized society. However, based on the quantum principle of inseparability, I'd argue that the opposite is true: when we all work together as a whole, our efforts are no longer fragmented, but congruent, and the results can transcend simple linear expectations.

This concept appears increasingly in the work of social critics. For example, the ensuing shift toward a collaborative culture that transcends extreme individualism is a major theme in Jeremy Rifkin's book *The Zero Marginal Cost Society*.[17] And in *The Ecological Thought*, Timothy Morton proposes that "interconnectedness implies radical intimacy with other beings and with nature."[18]

Picture humans living and working with the same efficiency as an ant colony, which operates as an indivisible whole. *Swarm theory*, also known as *swarm intelligence*, proposes that while the intelligence of an individual ant is minimal, the collective intelligence of the colony is remarkable. The collective engages complexity far beyond the ability of the individual to do so.[19] Here is a circumstance in which we can see that the whole is far greater than the sum of the parts. As a collective, the ant colony is superefficient. Indeed, businesses and organizations study and adapt the intricacies of swarm theory to solve extremely complex challenges. Imagine what embracing this cooperative spirit could do for the human race if we engaged the complexity of our challenges—poverty, climate change, disease, warfare, terrorism—with an equivalent complexity of intelligence. The cooperative spirit engendered by entanglement removes the divisive tendencies of excess individualism and leads to a sense of concert.

Let's consider the word *concert*. It speaks not only to symmetry, but also to a sense of harmony. Think of an orchestra. The individual musicians and their instruments don't lose their unique contribution, but they blend into an orchestrated concert so that the whole is not only greater than the parts, but is also immeasurably more complex and beautiful. Individual musicians suffer no loss of talent and uniqueness by merging into inseparability. Actually, one's particular, unique sense of self and identity is enriched, not blurred, by opening more fully to others. Our relationships with others and our own self can move into a nondisruptive sense of concert when we invite inseparability. Separation provokes disharmony and disconcert, while inseparability moves us into the harmony of the whole.

We can gain a deeper, more coherent way of knowing, an inner wisdom, when we begin to think in wholeness. The participatory worldview depicts an unfolding reality in which all parts, including us, share in a dance of unimaginable coherence. All we need to do to join in the dance is recognize that we are already participating in it.

BECOMING THE MASTER
OF YOUR THINKING

Welcoming the participatory paradigm into your life requires learning to think in a participatory way. This is a pivotal step in accessing your possibilities and freeing yourself from the limitations of mechanistic thinking.

Ordinarily, we don't pause to consider our rapport with our thoughts; they seem to operate autonomously and automatically, and so they script our lives. This chapter will reveal how you can learn to become the master of your thinking as you integrate the quantum principles of inseparability, uncertainty, and potentiality into your thinking process. We'll look at how our thoughts operate, and you'll learn to think in wholeness—in alignment with the participatory paradigm. Doing this allows a deeper and more profound sense of self to arise and evolve, one in which you're more than the simple aggregation of the millions of thoughts and feelings you've experienced.

THE POWER OF OUR THOUGHTS

To get a sense of what a major role your thoughts play in your daily life, to the extent that they will predict your future experience, just contemplate this question: Are your thoughts your best ally, your worst critic, or somewhere in between?

The most intimate and impactful relationship you will ever have is not with your parents, your partner, or your children. It is with your thoughts. They are your constant companions. Your thoughts will not only affect you far more than any other relationship, but will also greatly inform all your relationships—both with yourself and with others.

In recent years we've seen a flurry of attention surrounding the power of thought. The classic biomedical understanding of the brain proposes that our thoughts are affected by our brain chemistry and are a result of electrical and chemical processes. Many renowned thinkers, including the Dalai Lama and the late physicist David Bohm, have suggested it might well be the other way around—that our thoughts influence our biochemistry.[1] The Dalai Lama was influential in research in which the brain chemistry of deep meditators demonstrated that their brains were altered by their practice.[2] Indeed, emerging evidence from the field of neuroscience confirms that thought itself has the power to sculpt our brains.[3] We should see this as encouraging news, because if thought plays such a prominent role, then we are not at the mercy of our biochemistry.

However, thoughts can either be our supportive benefactors or our petulant antagonists. A particular thought—embedded as part of a dominant belief—can either imprison or empower us. In chapter 2, I explained how certain wave collapses we experienced set up confining primary beliefs about ourselves, from which a torrent of thoughts flows. But our thoughts can also flow from the positive wave collapses that earmark our defining moments.

We tend to believe that modifying our thoughts requires modifying our biochemistry with medication, yet that approach creates a multitude of side effects and treats only the symptoms. Seeking a quick fix to the problems caused by our thoughts also robs us of the opportunity to develop a healthy relationship with them. If negative thought impairs our brain chemistry, then healthy thinking improves it.

Neuroplasticity is the brain's ability to reorganize itself by forming new neural connections throughout our lives. The very notion indicates that the brain is flexible to a greater extent than previously believed. Rather than thinking of your brain as the factory that manufactures your thoughts—mechanism at work, once again—consider it in the following way: If you're walking on the beach and look behind you, you'll see your footprints in the sand. The sand didn't manufacture each footprint; your foot simply left its imprint there. In the same way, thought leaves its mark on the brain. This realization should open a pivotal shift in our operating beliefs such that we no longer view ourselves as "hardwired" in a certain way or the victim of our brain chemistry, but as fully capable of creating a self-actualizing experience.

To further this assertion that our brains don't produce our thinking, consider the phenomenon of near-death experience. In countless circumstances, people have been declared brain-dead as their hearts stop and their brain-wave activity flatlines. Although they are presumably dead, after a period of time their vital signs return to normal, and in many cases they are able to repeat the conversations and describe the circumstances that occurred after they appeared to be dead. Consciousness evidently survived the death of the brain.[4]

When we marginalize anomalies such as these, we perpetrate bad science. Even just one exception to the rule—let alone countless exceptions—should demand a rethinking of what we hold to be true. There is ample evidence that in many cases consciousness does not require a functioning brain.

THE LANGUAGE OF MECHANISM

Our commonly used words, terms, and expressions tell us how we've learned to see ourselves. How often have you heard the term *hardwired* in relation to our brains? The term "hardwired" corresponds with the mechanistic worldview, wires being part of a machine's apparatus. Humans have no wires, but when we begin to ascribe such features to ourselves, we can see the mechanistic paradigm asserting its imprint.

A young man with whom I'd been working proclaimed, "I think I have a screw loose." Humans don't have screws; machines do. Equipment may be dysfunctional; people cannot be. A couple may ask me if their marriage is "reparable," but a relationship is not comparable to a machine that needs to be repaired.

The very fact that we apply the words *function* and *repair* to people or relationships speaks loudly of the mechanistic approach. The word that I'll use to describe this characteristic—imputing machinelike qualities to humans—is *mechanomorphism*. This is the unconscious process by which we take parts of the machine and apply them to our very being. Consider the damage we do to ourselves when we reduce the unique qualities of being human to the bits and parts of a machine. The human body and mind consist of an incomprehensible information network that correlates all parts at warp speed—hardly a machine with wires.

When our words and our thoughts remain rooted in the language of mechanism, we are imprisoned by that picture of reality. Shifting our worldview allows our thoughts to escape the bondage of mechanism. As we do so, we will open to thinking in a participatory way that allows us to see in wholeness rather than parts.

CHANGING OUR THINKING VERSUS CHANGING OUR THOUGHTS

What we have been discussing about the ability of thoughts to change our brain chemistry appears to support what a number of well-intentioned pundits in the self-help field would have us believe: that the solution to this problem is simply to change our thoughts. Their basic maxim is "Change your thoughts; change your life." (Indeed, half a dozen books have appeared in just the last few years with some variation of that phrase in their titles.)

Yet, in and of itself, this exhortation is insufficient. We don't want merely to substitute a positive or preferable thought for a negative one because that doesn't address the more important matter of how we think. *How* we think is ultimately far more important that *what* we think. What I am proposing is a new mindscape in which our thinking flows from the coherence and wholeness of the participatory paradigm rather than from separation and the fragments that result from the mechanistic worldview.

The moment we have a thought, we don't actually see the thought operating, and so we become attached to that thought. We become the thought and automatically summon our personal biography of accompanying emotions, thus reacting from conditioned reflex born of the past. These ensuing emotions then trigger another corresponding thought. This is why we cycle up or down emotionally as we follow the roller coaster of our thoughts and feelings. Thought and feeling act in tandem, and I find thought to be the pivotal factor in this relationship.

I've previously said that our thoughts are primarily informed by our beliefs, many of which are the results of our experiences (wave collapses). I'll go further and say that thought tends to represent our beliefs and life experiences. To *represent* should be read literally as to *re-present* because we often get stuck in deep basins of old, habitual thought as it continues to present our past over and over. Consider it this way: thinking is constantly active and new, whereas thought tends to be old and habitual.

Thinking is the ability to see your thoughts operating. Our objective is to seek new thinking, which frees us from the replay of old thoughts.

LEAVING THE FAMILIAR ZONE

Breaking free of old thought requires coming out of what I call "the familiar zone." Ordinarily, we might call the realm of old thought the comfort zone, except that it's often not particularly comfortable—simply familiar. If you suffer from self-critical thought or low self-esteem, you know exactly what I'm speaking of. The struggle to liberate yourself from the confinement of old thought and to engage change, if not transformation, requires surging beyond the boundary of the familiar zone again and again.

When I'm working with individuals to help free them of old thought, they may say, "It's too scary" or "It makes me uncomfortable." They may also proclaim, "I don't know if I can do that."

This tendency to defend old thought came up in a session with my client Josephine. My goal was to assist her in breaking free from her avalanche of negative self-talk. This habit blocked her ability to advocate for herself with her husband when they were in disagreement or conflict. But when I proposed some new communication skills, she protested saying, "That's not easy to do." I asked her how she knew that to be so if she had never tried it. Upon reflection, she admitted that she couldn't really know that; her response was automatic—old thought defending its territory.

Based on years of experience, I've determined that these excuses and justifications are part of the complex mental defense mechanism that we generate to guard against new thinking. Often, any threat of movement beyond the bounds of the familiar induces stress or anxiety. And yet the avoidance of this discomfort is precisely what keeps us stuck in the familiar zone. As we saw from Angela's case in chapter 2, as miserable as she was in her marriage—her familiar zone—she froze with fear at the thought of breaking free of the familiar.

The familiar is exactly what has lulled us into sleepwalking for so much of our lives. (I'm not referring to our familiarity with the things we may cherish—love, loyalty, friendship, stability, and so on. I'm talking about the areas of our lives we seek to change.) And our relationship with the unknown is thoroughly informed by our indoctrination toward seeking certainty and predictability.

TURNING DISCOMFORT INTO YOUR ALLY

One way I help clients move beyond the self-imposed boundaries of the known is by encouraging them to shift their relationship with their discomfort. If the anxiety of moving into new terrain becomes our justification for maintaining old thought and behavior, we need to alter our relationship with this anxiety. In chapter 2, I proposed that we need to change our relationship with uncertainty; in this case, it's the uncertainty of new thinking that we need to change our relationship with.

Instead of letting discomfort immobilize you, look at the discomfort as your ally, as a signal that you're venturing out of the familiar zone. Turn the disquiet into your personal coach as it propels you forward. If you're comfortable, you're probably stuck in the habit of old thought.

Just as working out physically and training our bodies requires physical stress, learning to free our thoughts from their habitual grip on us usually requires some level of psychological stress. Countless people readily engage the discomfort of a challenging physical work-out in return for the anticipated benefit to both health and vanity, yet we're more tentative in surging beyond our familiar zone on emotional and psychological levels. If we put as much effort into improving our internal processes as we do to improving our physical condition, we'd be surprised at how accessible change becomes.

I recall a walking therapy session I had with a young man at the beach not far from my office. Mark was burdened with social anxi-ety, and he self-protected by acting reclusively, leaving his home only to attend college classes. Just considering coming out of his familiar zone and socializing with peers overwhelmed him and provoked acute stress and anxiety. Rather than engaging his fears so as to allow his personal growth, he avoided this distress altogether. He found himself in a double bind as he used his anxiety as a justification for not break-ing out of his zone; yet the consequences were that he was living an isolated and depressed life.

I picked up a stick and drew a circle in the sand around Mark. He asked why I was doing that. I asked him to remain within that circle while I walked off to get us some lunch. He seemed disconcerted and asked if I was serious. I smiled as I explained to him that the circle I drew around him represented how he had constrained his life.

Mark was remaining confined within his familiar zone, with results that were deleterious to his life. As a rule, remaining stuck in the famil-iar zone contributes to feelings of depression as it denies us the vitality of new experiences. But if Mark could embrace the anxiety, discomfort, and uncertainty that accompany leaving the familiar zone, he could begin to disable it. Once again we see how embracing uncertainty—in Mark's case, uncertainty about socializing—can move us beyond the stagnation of our familiar zone and open us to new possibilities.

Having embarked on a path beyond the bounds of the familiar, we may rest in the satisfaction that we've grown. As we begin to expand the circle that confines us, it may become somewhat larger, and so we may pause in our expansion and breathe a sigh of relief. Yet we will still have an inclination to be pulled back toward the familiar zone, as if it had a gravitational field and we were in its orbit. This tendency persists not only in our struggle with individual growth, but also in the expansions and contractions that occur in our relationships. The way to counteract this tendency is to keep expanding farther and farther into new territory by continuing to embrace the new possibilities offered outside the familiar zone. As we do, our familiar zone grows immeasurably larger, and we eventually break free of the old orbit.

When I encourage people to embrace their discomfort, they often speak about their fear of the unknown. Instead of fearing the unknown, it's helpful to reenvision it as freedom from the known. The known is where certainty resides, and at times this is exactly what we're seeking to change. Although this known has become familiar, it's where we get stuck. By embracing uncertainty, we can break free from the confines of the known into new possibilities.

SHIFTING FROM HAVING THOUGHTS TO THINKING

Those of us old enough to remember vinyl records might recall that when there was a scratch on the album, the needle would sometimes get stuck in a groove. The same music or lyric would keep repeating, and with the blockage, the tone arm couldn't find its way into the next groove. Likewise, our thoughts tend to keep reiterating the same messages ad nauseam. As they do, they summon old memories and feelings, and we struggle to change.

Thought is automatic in that it presents itself without our noticing it. We therefore become trapped in a rut of old thought. The first step in liberating yourself from this mental groove is learning to *see* or notice your thoughts. If we don't notice our thoughts, we attach to them and are reduced to being the thoughts. Developing an awareness of thought, what I call thinking, allows you to remain above the thought as you observe it.

Seeing our thoughts is a matter of timing. With practice, we can learn to become more alert and see an individual thought operating.

This process of becoming alert to our thoughts and what they are doing is a little like watching a slow-motion replay from a sports event: you see the play unfolding slowly and clearly. We are truly thinking when we can see our thoughts operating. That's the goal of thinking.

For another analogy, let's look at tennis. The tennis ball is our thought, and becoming aware of our thought is like anticipating the arrival of the ball on our side of the net. We see our opponent's positioning and footwork, her racquet movement, and the ball as it advances toward her. By the time she hits the ball and it approaches the net on its way back to us, we're fully engaged and anticipating its arrival. We'd hardly wait until the ball was inches from us to react. Anticipation and awareness are fundamental in tennis—or in any sport, for that matter. We can train ourselves to be aware of what's happening to the ball at all times, and time slows in a relative sense as we come into this zone of awareness.

The same thing can be accomplished with thought: we can learn to *see* it in advance of *becoming* it. Developing this awareness of thought is essential to achieving a mastery of thinking. If I can't see the thought, I won't be having a thought; the thought will be having me.

Judging your thoughts gets in the way of simply observing them. If you judge them, it's still your old thought operating.

Throughout your day, try to notice your thoughts. Imagine sitting in front of a large TV monitor and watching your thoughts transcribed on the screen. Don't judge them; just see them. Just sit back and, in a detached way, observe what you're seeing.

As you develop the acuity to see your thoughts, you'll be creating an important tool toward your mastery of thinking. You're developing a powerful muscle memory—the ability to witness your thoughts.

Once you have developed your ability to notice your thoughts, you can begin to imagine an old thought as a visitor knocking at the door of your attention. You may hear the knock, but you can decide not to open the door.

Our old thoughts come at us with tenacity. If you find a particular thought won't stop knocking, try the following technique, which many of my clients have found helpful.

> When you notice the old thought clamoring for your attention, place your forefinger vertically in front of your lips and say "shhhhh" to the thought! Remember that you can choose not to open the door to it. The thought will continue to try to draw your attention, so be persistent.

As you progress in your ability to witness your thoughts, you can look at the recurring themes that they present. You'll likely notice a few central topics that your thoughts are drawn to. There may be numerous variations on this theme. The core belief that "I always choose the wrong friends" may include thoughts that range from "What's wrong with me? Why didn't I see that she was so selfish?" to "Other people seem to have loyal people in their lives. Why don't I?"

These recurring themes tend to be limiting and often become self-fulfilling prophecies, as we've seen. They are often constellated around our confining wave collapses and limit our possibilities. It's essential that we notice how they attract our attention like a magnet. To break free of their pull and achieve a mastery of your thinking, you must first become aware of the tug of these thoughts. Try tracking these thoughts back to their source, which is your fundamental beliefs about yourself.

Once you've progressed in your new ability to see your thoughts operating, you will notice a space between your thoughts. Within this space is the extra moment required for reflection. This space is where your possibilities reside. You can engage the state of superposition that I described in chapter 3—your state of potential. In the instant before you become your next thought, everything is possible. A primary tenet of the Possibility Principle is that this is the moment for us to become aware of.

Developing a heightened level of awareness regarding our thoughts permits us to be responsive rather than reactive. Between the instantaneous reactions of our thoughts lies a wiser and deeper sense of self, poised to emerge. This self notices its thoughts (and ensuing feelings) and isn't reduced to simply being their end product.

In the space between my last thought and my next thought, I emerge. The "I" to which I'm referring is wiser and deeper than my mere thoughts. I can choose to think differently, to see things differently. This moment of awareness brings us to the precipice, where we can elect to become the master of our cognitive process and herald defining moments. These are new defining wave collapses in which we break free from the confines of earlier, confining ones. We don't know what's to come, but we cultivate the space for new thinking to emerge. It bubbles up when we can get out of our own way and allows our deeper intuitive wisdom to arise as we develop a relationship with our deeper being. It enables us to come out of our script, to break free from our stories in which our old wave collapses trigger tired beliefs and thoughts.

LITERAL THOUGHT VERSUS PARTICIPATORY THINKING

Remember that many of our recurring thoughts are the results of confining wave collapses; when we are unaware of where they are coming from, we see them not as mere thoughts but as the "truth." If you have carried a thought such as "I'm not good enough or smart enough" or "I'm a failure"—or in the case of Jill, "I'm not lovable," or Helen's "I'm not important"—you don't see it as a thought; you blindly accept it as objective truth. These confining thoughts form your primary beliefs and script your experiences and life dramas.

Not recognizing that our thoughts are falsely reporting a truth to us, we accept the story they are telling us. This is known as literal thought. Literal thought reports in on the so-called truth that it is seeing; it may sound like "You are . . ." or "I am . . ." or "It is . . ." This manner of thinking is completely at odds with the participatory paradigm. From the participatory worldview, our subjective thoughts and perceptions participate in our reality-making process.

To understand the difference between literal and participatory thinking, let's go back to the example of Josephine. As I was encouraging her to break free from old thought and she protested, "That's not easy to do," her thought asserted what appeared to be a truth. Her literal thought was stating as fact something that was actually a belief or perception. What I encouraged her to say instead was, "I noticed that my thought was telling me this was going to be hard to do." Josephine could then distinguish *what her thought was telling her*. This rewording was far more than simple semantics; it moved Josephine from feeling like a victim of circumstances toward being the creator of her experiences.

If we can separate from the cognition and notice what it's saying, we are more than purveyors of old thought; we are thinking. We are taking charge of our thinking process and inviting in the possibilities that new thinking affords.

As we've seen, this observation also slows down the automatic nature of our thought, allowing us to enter into the realm of thinking. Elite athletes often claim that when they're at their peak of performance—"in the zone," as they say—they seem to see the action in slow motion. This experience is what we should seek as we slow down the torrent of old thought processes. This higher state of awareness allows us to *see and choose our thoughts* and to develop a deeper and more grounded sense of ourselves that is more sovereign than the thought. It allows us to find the wiser and more genuine self that has been submerged beneath the clamor of our thoughts.

In working with my client Jane, she acknowledged her compulsion to keep her children peaceful and quiet at all times. Her recurring thought was, "I need to quiet down the kids," even when there was no reason to. As we worked on her ability to see her thought and notice what it was telling her, she was able to appreciate both the habitual nature of her thought and, just as importantly, how it stated as a fact what was really a compulsion coming from somewhere else. She was having a recurring thought that told her she needed to keep the children quiet, but was that the truth? *Why* did she need to always keep them quiet? Where did that message come from? We discovered that this message was a carryover from what she had heard in her own childhood. Once she realized where her recurring thought was coming from, she could choose to break from her operating belief and think differently.

Consider the difference between the following statements:

1. **"He is so selfish and cares only about himself; I mean nothing to him."**

2. **"I'm having a thought—or feeling—that is telling me he's uncaring and selfish."**

In the first statement, my thought is asserting an unyielding truth. It speaks to an objective reality about "him." But you might ask yourself if this is true for him in other relationships. Do others see him differently? And how might your actions influence his behavior? Your role in that thought is entirely removed; you have no participation with the thought, which is proclaiming the "truth." This is literal thought at work.

The second statement reflects participatory thinking and takes note of what the thought is saying.

You may wonder, what's the big difference? The first (literal) thought is objective, drawing a conclusion that leaves no room for further contemplation. The second (participatory) thinking is subjective. It's about how you see him. Perhaps you see him that way because you said or did something that made him reactive or mean. Or maybe he has a terrible headache or is stressed out. Or perhaps he just doesn't like you. Participatory thinking makes subjective representations, while literal thought comes to objective conclusions.

Whether you're communicating with someone else or simply reflecting on your own perception, the preface "I'm having a thought, which is telling me . . ." severs the automatic tendency of your thought to assert its position. It enables you to take ownership of the thought and see it operating. I cannot overstate how essential this is in initiating your mastery of thinking. Ultimately, you can come to see that your thought is just a thought. Using this prefacing technique gives you the moment to apprehend the thought and see what it's telling you. Then the thought need not become your truth. As we'll see in chapter 11 on communication, this approach is instrumental in getting past the mind-numbing roadblock of conflicted relationships.

From the participatory worldview, separation falls away, and all parts engage in an inseparable, flowing reality, of which our mind is a primary participant. From this vantage, we can understand thought as subjectively representing what we *think* we see out there. Because our consciousness participates in that observation—recall that the uncertainty principle informs us that we are affecting what we observe—we are no longer detached observers or victims of the circumstances. It's not happening to us so much as we are participating in the happening. We are orchestrating what we participate in.

Just as seeing your thought allows you not to become your thought, the same holds true with your feelings. The goal is to see your feelings arise, identify, and note them. If you're feeling hurt or angry with another person, express what you're feeling instead of becoming the feeling. In a couple's session, Claire's voice became angry, and her body stiffened as she recounted how her husband had failed her as a co-parent. Her hostile communication could only cause Jay to defend himself. I asked Claire to pause, notice what she was feeling, and express the feeling rather than becoming the feeling.

On an intrapersonal level many people become consumed with the negative feelings that correlate to their habitual thoughts. If you have depressing thoughts, you'll probably feel depressed. The thought is the linchpin, so it should be the pivotal focus of our aim to free ourselves from the groove of depression. If I'm struggling with a troubling feeling, I ask myself what thought I've recently attached to that might have triggered the feeling. I can usually track back and find the thought and then release the thought-feeling complex that is affecting me negatively.

I was working with Mateo on his core self-esteem issues. Because of his poor self-image, he worried relentlessly about what others thought of him. He measured his words and actions as he tried to calculate how to induce others to like him or approve of him. I helped him look at his recurring question, "What will they think of me if I say something that sounds weird to them?" From this literal-thought stance, Mateo believed that others were indeed judging him, evaluating his words and actions. He judged himself, projected that judgment onto others, and then imagined he was on stage for others to critique. After teaching Mateo the difference between literal thought and participatory thinking, I asked him to express himself in a participatory way. He said, "I keep having the same worry, the same old thought that others are judging me. But that's just what my thought is telling me and most likely not valid."

Literal thought has us project our subjective belief onto others and ourselves and ascribe an objective reality to it. Try this exercise.

The next time you experience an upsetting feeling, pause and ask yourself what literal thought is setting up that feeling. You can see the literal nature of that thought as it pretends to be an objective truth, such as "He never . . .," "She always . . .," They are . . .," or "I am . . . "

Now take the literal thought and rephrase it in a participatory way by saying, "I'm having a thought that is telling me . . ." You can now see your participation in that thought and ensuing feeling.

When we see a particular negative thought often enough and begin not attaching to it, it retreats, allowing new thinking to arise. We develop a new muscle memory, so to speak, and become free of the past. You may recall the story I told in chapter 3 about Helen, who was certain that the person she was supposed to meet had blown her off because "I just wasn't important enough." That was a clear case of literal thought—coming from Helen's confining wave collapse of being treated as unimportant by her mother—telling her the "truth" about the circumstances. I trust that if something similar occurred now for Helen, she'd be free of her imprisonment to literal thought and the confining belief from her past.

When we ask ourselves how we know something to be true, most often we will admit that we can't be certain but that it's probably based on either our limited personal experience or our belief. Embracing the uncertainty of not knowing for sure is not only advantageous on a personal level, but also resembles the emerging quantum worldview. The classical paradigm predicated on certainty sets up the trickery of mind that equates thought with the truth. This false sense of certainty then impinges on new learning and self-reflection. Uncertainty opens the doorway for new possibilities and new levels of inquiry. This is a dynamic experience that liberates us from the prison of habitual thoughts and beliefs.

Ask yourself, "What if my beliefs about myself or others aren't necessarily true?" Although it may be daunting to admit that our life stories are subjectively constructed, doing this frees our energy to reconsider the stories we've constructed, and we can then prepare to alter the script.

6

MOVING BEYOND EITHER-OR

The truncated way we see reality has been influenced in large part by what is known as Aristotelian thinking. Aristotle believed that reason is humanity's highest faculty, and his theory of logic accounts for what today we call *deductive reasoning*, which links premises with conclusions. Briefly, if your premises are true and you follow the rules of logic, then the conclusion you reach must be true. The corollary of Aristotle's philosophy is that things either *are* or *are not* true. There is no gray area, no middle ground. Aristotle had a profound influence on the philosophers who followed, including Descartes, and some of his ideas on physical principles became part of Newton's laws. The duality that resulted from Aristotle's theory of logic filters how we picture reality operating and is known as *either-or thinking*. It structures our beliefs into an oversimplified posture by which we know something only by including its opposite. This way of organizing reality frames our thought process as it takes the undivided wholeness of reality and fractures it into two opposing camps.

The word *hate* wouldn't have much meaning without the notion of *love*. *Good* wouldn't make sense without including *bad*. The pairing of opposites helps us differentiate things, and these notions shape the mindscape of our reality. But it also entrains our thinking to look at the opposition and forget that there is an included middle—a place where the opposites can coexist.

This dichotomy of either-or thinking naturally leads us down the path to being either right or wrong. Our identity becomes wed to what we believe to be the truth, once again rooted in the literal thought of the classical worldview. Our thoughts' need to be right works overtime to protect this self-constructed image of our identity. Most people protect their egos by defending the need to be right. The more rigid our thinking, the more fragile is our sense of self. It also follows that the more tied we are to being right, the more fixed our beliefs and thoughts become. The absence of flexibility in our thinking suggests that we're more interested in being right, in winning, than in learning or relating. This is another form of thought defending its territory and doing battle against the possibility of being wrong.

The paradox here is that vulnerability of thought—being comfortable with being uncertain or wrong—actually makes us powerful and strong. Once we free ourselves from the *need* to be right, our thinking and our discourse open up. Not being tied to defending the need to be right opens us to wonder, inquiry, and new possibilities. It also provides an exceptional nutrient for our relationships.

What I'm proposing is that we reconsider vulnerability as we ordinarily think of it, which is feeling at risk or insecure, and to consider that it's just the opposite. If I don't need to be right, I'm not embarrassed to be wrong, which makes me far more powerful. If I have nothing to fear with regard to what others think, my sense of self isn't tied to such a silly artifact, which fosters a more powerful self-esteem. When you listen to an interview and the guest responds to a query by saying, "That's a good question," it suggests they have a ready answer. When I'm asked a question that provokes my being uncertain as to the answer, I relish it. It opens the door for me to dive into the uncertainty and see where I'll come out. That's a learning opportunity for me.

Ask yourself in what ways your need to be right might block your ability to connect to and validate others or to open yourself to new learning. Try going against the grain by suspending the need to be right and see what happens.

TRANSCENDING DUALITY

In the quantum realm, we looked at the wave-particle duality in which light can be both a particle and a wave, demonstrating that the very notion of either-or becomes unsustainable as a new vista of reality is introduced, one in which things *are* and *are not* at the same time. This phenomenon is just as true in our everyday lives. Our rational desire to break things down into neat, distinct categories no longer helps us function. In fact, this oversimplifying may contribute to many of the difficulties we face. As I described in chapter 4, our inclination to fragment creates havoc or crisis, yet we are blocked from seeing our part in it. We need to engage a higher level of complexity—beyond either-or thinking—to meet the multifaceted challenges that we face.

We can see how stuck things become when we are rooted in the duality of either-or thinking. Pay close attention to how questions are posed, and you'll notice how questions often rupture and fragment wholeness as we slice and dice reality into distinct compartments. For example, at twilight, is it day or night? This question provides only two choices, which obscures a more nuanced response. It's really day blending into night. Do you love me or hate me? That might depend on the moment—although either term might be overstated. Do you believe in creative intelligence or in evolution? This sets up a simple one-sided answer. Is it not possible to believe in both, or does one preclude the other? Can I be pro-choice yet highly sensitive to the ethical issues of abortion? I would like to think so. If your partner is upset with you and asks you to acknowledge how he or she feels, can you both validate your partner and not default into necessarily being wrong? When I teach my Mastery of Thinking course, someone invariably asks, "Mel, do we discover reality or create reality?" The either-or question fractures reality into two distinct compartments. My quizzical response is simply *yes*. I have retrained my mind not to fall into either-or thinking. I believe that we both discover and create at the same time. One doesn't preclude the other.

Discover suggests that reality lies "out there," while *create* implies that the source is from within. The very demarcation between *in here* and *out there* is merely a way in which we've been trained to see reality. From a participatory worldview, we can appreciate that such distinctions become indistinct. Did Einstein discover the theory of relativity or did he envision it? The new science suggests that this duality is simply our way of picturing reality—not how things actually operate. When we transcend the limitations of either-or patterns, we begin to think and see in wholeness. We can then reverse the thinking that had ruptured wholeness into separate compartments. This new thinking is participatory and moves beyond the either-or divide and is of immense help in our relationships and communications, as we'll see in coming chapters.

Shifting from *either-or* to *either-and-or* enables us to engage complexity and resist the temptation to oversimplify and fragment. Complexity, which includes both opposing parts, permits all opinions and positions to be heard and considered. It does not suggest, however, that we'll remain mired in indecisiveness or lacking in clarity. A deeper and more authentic simplicity emerges from complexity. Oliver Wendell Holmes Jr. wrote, "The only simplicity for which I would give a straw is that which is on the other side of the complex—not that which never has divined it."[1] He is suggesting that we resist the black-or-white thinking that entrenches our positions and embrace the shades of gray that will ultimately inform our deeper and more educated beliefs. When we do, we break free from the straitjacket of the right-or-wrong dilemma.

As a society, we are trained to avoid complexity and especially confusion. We desire simple, neatly divided answers and solutions—quick fixes. Reality appears pluralistic (containing innumerable qualities), nuanced, and highly subjective. Trying to extract simple right-or-wrong, yes-or-no answers from such a tapestry is wrong-minded and foolish, if not dangerous. It is incoherent and not much different from trying to place a square peg in a round hole. It doesn't fit, just as either-or thinking, for the most part, is incongruent with a participatory reality.

KEEP YOUR THINKING VIBRATING

I recall reading a number of years ago that if concrete is kept vibrating, it won't set but will retain a liquid form. This concept intrigued me, and I considered it a great metaphor for how we see our thinking. Keeping our thinking vibrating keeps us from falling prey to our personal dogma or to *concretizing* our beliefs. To accomplish this requires a flexibility that allows us to avoid the pitfalls of either-or thinking.

Embracing confusion and complexity increases the bandwidth of our brainpower while stimulating and generating far more productive, flexible thinking. Oversimplification—to which our society is grossly addicted—makes our minds rigid and inhospitable to genuine thinking, keeping us stuck in the confinement of old thought. By embracing uncertainty and actually inviting confusion, we keep the vibrating alive and also increase our neuroplasticity, the workout for our brain's flexibility. Inviting confusion—normally shunned by the dictate of certainty—opens the gateway for generative thinking.

The absence of mental vibration predisposes us to get stuck in a virtual groove of thought and experience. From such a fixed state, we struggle with change. Vibrating enables a perpetual questioning, but not from a point of insecurity, doubt, or analysis. The goal is to embrace an open-minded inclination to wonder to oneself. A relationship with inquiry paves the way for such oscillating. Genuinely inquiring minds seek new questions rather than longing for answers. The answer ends the inquiry, and the vibration halts. Reaching temporary and relative answers for a while is fine—a resting place, if you will—but the inquiring process should continue. The goal is to find a state of equilibrium between knowing and not knowing, confusion and clarity.

Engaging paradox stretches the mind beyond the limits of everyday logic and rationality that constrain us. Reflecting, contemplating, and looking at things from multiple perspectives enables our minds to be flexible and creative. This keeps us oscillating. This ongoing state of uncertainty, in which we learn to suspend our assumptions and beliefs and to reevaluate them continually, becomes the groundswell for evolving insights. Because reality presents itself as an inexorable flow, we want our thinking to inexorably flow as well, embracing uncertainty to avoid a fixed state.

ASKING NEW QUESTIONS
OPENS NEW POSSIBILITIES

The participatory worldview demands new types of questions. If we continue to ask the same types of questions, searching for simple answers rooted in either-or responses, we resist entering uncertainty and complexity. Questions derived from classical mechanistic thinking will continue to lead to answers based on that outmoded worldview. Only questions based on the participatory worldview have the potential to shift our attention in important ways. These questions invite contemplative answers that don't reduce us to yes or no, right or wrong.

If our questions become more imaginative and evoke curiosity or wonder, we can participate in a potential-laden, unfolding reality in which the answer is not necessarily "out there" but rather both in here and out there. With this approach, the question and the answer become complementary aspects, similar to the wave and the particle. The way the question is asked shapes the answer. In this way, questions are far more powerful than answers, as they direct where our attention goes.

Within interpersonal relationships, for example, asking new or different questions makes us present to new engagements and opportunities. Relationships become stale, if not deadened, as old questions mandate old responses. No one need be truly present as we sleepwalk through the conversation.

Such a new question came up for me in a couple's session I was facilitating. As François consistently demanded change from his girlfriend, Patty, they argued back and forth about who was at fault and who was right or wrong. It occurred to me that we were missing the bigger question. I wondered if Patty had any desire to be the kind of girlfriend François was asking her to be. I prompted him to say this to her: "Forget about what I'm asking you to change. What changes do you wish for in yourself?" I had never heard that question asked, but I consider it fundamental to our requests for change in relationships.

New questions can be provocative and generative. They are essential in creating resilience in relationships, lest our relationships suffer from predictability. Follow the path of a unique inquiry, and uncertainty abounds. There lies the excitement of genuine engagement and dynamic learning.

WHAT INFORMS YOUR BELIEF?

While I was delivering a somewhat provocative talk on the subject of change, a gentleman in the audience indicated that he had a question. As he began to speak, it was evident from both his tone and his question that he was challenging the material that I was sharing. Simply stated, his core belief was that people don't change, and he suggested that I was being an idealist. Little could he have imagined that I welcome the charge of idealism, for this is what inspires us to higher levels. So I caught him by surprise when I thanked him for his compliment. Then I asked him where the world would be without the benefit of ideals.

Nevertheless, his tone remained charged as he continued to assert his position. My presentation was evidently offending his beliefs. I felt my own instinctual drive to prove him wrong and reveal the flaw in his thinking. Thankfully, I was able to see my reaction and not become it, and so I resisted the need to be right. In that instant, I chose not to default into the adversarial match of right and wrong, either-or thinking—either people can change or not. Instead, I witnessed my emotion, quieted myself, and came into a space that permitted a more authentic response and prompted more meaningful discourse. Within a moment or two, I felt a question arising from within. It percolated from a deeper place and took form in these words: "May I ask what informs your belief?" It was a question that I had never previously asked.

Not surprisingly, my question took him off guard, and he struggled with his response. Since our thoughts are underscored by our beliefs, understanding his belief was important to engaging him in dialogue and relating with him. Where his belief came from was far more important to me than what his belief was. I was asking a question intended to open up a deeper inquiry, rather than shut it down. The man, who never gave his name, eventually suggested that his belief was caused by his personal life experience. He offered that he had never been successful in changing any major aspect of his life. Now we were on to something. I shared my personal success with the change process and asked him if he thought he'd have a different belief if he had had my life experiences. He reluctantly allowed that he might. The doorway now opened toward an evolving conversation.

Asking someone what informs their beliefs is both respectful and genuine. Without such an inquiry, we're simply asserting our literal thoughts at them. For example, when people utter sexist or racist comments, I find it utterly futile to try to refute their statements. Asking instead how they came to believe what they do is much more enlightening. They might share some deeply held beliefs or personal life experiences, which then could open the pathway for a deeper understanding of how they came to their beliefs.

When we express our positions in a participatory way, by subjectively sharing what informs our beliefs, we invite a participatory discussion and, hopefully, a more vital communication. This manner of discourse is far removed from the so-called objective (Newtonian) statements of fact that degenerate into neither party actually listening. This style of communication plagues us at all levels of relationship—from personal intimate relationships to opposing experts and pundits all the way to the level of international and religious conflict. After all, war and terrorism are the manifestation of conflicting beliefs that are typically not exposed to a shared inquiry. Engaging in a shared inquiry upends the penchant for antagonism and conflict.

Most of us become deeply identified with our core beliefs. When this occurs, we struggle to separate our beliefs from our identity as they coalesce, and we defend our beliefs and our identity mightily. When communicating with others, we usually don't share how we came to these positions, and in not doing so, we cut the conversation off from the flow of vital information. What follows is typically frustrating because neither party is open to reflection or new learning. On a more intrapersonal level, appreciating how we have become influenced to think and believe as we do opens pathways for change within us. Beliefs are informed and shaped by our family of origin, our culture, education, experiences, and worldview, to cite just a few factors. We shouldn't make the error of concluding that these are commensurate with being objective truth, for if they are, they will deny the truths of other people and make dialogue impossible.

Pick one personal belief that feels especially significant and ask yourself how you came to have it. Was it informed more by what your parents told you or how they acted, or by your personal experiences?

Once you see how you came to your belief, you can choose to reaffirm it or release it. The key is that you're in charge of what you believe, not the other way around.

By this point, I hope it's becoming clear that the very notion of objective truth is altogether doubtful in most cases. Inseparability and the indeterminism of uncertainty highlight the challenges to classical notions of objectivity. The next step is to understand how upending objectivity enables us to free ourselves from its constraints.

FREEING OURSELVES FROM THE GRIP OF PATHOLOGY

What I refer to as the "myth of objectivity" lies at the heart of many of our psychological and emotional challenges. Objectivity imposes its mandate on both the field of psychotherapy and our very notion of disorders and pathology. Even though the principles of inseparability and uncertainty render objectivity effectively meaningless, mainstream Western psychotherapy still remains largely rooted in objectivity. Notable exceptions to this rule have emerged in recent decades, but for the most part, current therapeutic practices ignore the discoveries of quantum physics.[1] Western therapy's devotion to diagnosis—which requires objectivity—blinds it to the primary cause of our malaise: the mechanistic worldview. And, as importantly, this outdated view victimizes us further as we construct a reality that sees pathology everywhere it looks.

EXPLORING THE MYTH OF OBJECTIVITY

We've seen that the discoveries from quantum physics uproot the very idea of objectivity. And yet if you think about how most people use the word *objectivity*, you will conclude that they see objectivity as a positive attribute, implying fairness or the absence of bias or prejudice. "You're not being objective" is taken as serious criticism. But peering into the actual semantic definition of this word, we can reconsider its plausibility from a nonscientific perspective. *Webster's Third New International Dictionary* (1993) defines *objective* as "of or relating to an object, phenomenon or condition in the realm of sensible experience, independent of thought and perceptible by all observers. Having reality independent of the mind; expressing or dealing with facts and conditions as perceived without distortion by personal feelings, prejudices or interpretations."

What does "sensible experience" mean? *Webster's* indicates that the word *sensible* refers not only to one's senses but also to matters of "reason or understanding." Yet the concept of being sensible is completely contextual and relative. We might agree that it's not sensible to run out of your house or apartment naked; that would seem bizarre and might expose you to a psychiatric evaluation. But what if you awakened to find your home on fire and had not a moment extra to escape? Under those conditions, would it not be sensible to flee your apartment without clothing (assuming that's how you sleep)? Most societies deem it illegal to murder another except in self-defense. Yet those rules change when a nation declares war and the context of killing another alters. Sensibility obviously varies based on the relative circumstances and changes as mores and customs shift. If circumstances and consensus determine sensibility, then sensibility is hardly objective.

The next part of the definition, "independent of thought," implies that an object, phenomenon, or condition exists in a separate and independent state that has nothing at all to do with thought. Without venturing into the philosophical and epistemological considerations this proposition evokes, how could we be sure something exists without thought? Not only does this premise obliterate the significance of thought, but it also refers us back to the concept of literal thought, which attributes a separate, superior existence to conditions and objects.

The absoluteness of "perceptible by all observers" leaves no room for exceptions. So if we now had one hundred or one thousand psychologists observing an individual, they would *all* have to concur without exception. Good luck! We can consider limitless circumstances in which what we consider an objective reality is not, in fact, perceptible by all. Witnesses to calamities or crimes often report seeing differing things. We know all too well the stories of people sent to prison based on eyewitness testimony, only to be found innocent years later through DNA evidence. Our perceptions inform what we see. Moreover, cultural differences greatly influence these perceptions, as what is normative in one culture may be bizarre in another. And so "perceptible by all observers" is an invalid concept.

The last part of the definition adds "having reality independent of the mind; expressing or dealing with facts and conditions as perceived without distortion by personal feelings, prejudices or interpretations." This definition of the word *objective* speaks directly and deeply to the root of the Newtonian-Cartesian paradigm, informing us that an independent, detached reality exists "out there," separate and distinct from us, and that our consciousness and perceptions have nothing at all to do with it. We've seen how the new science has decimated that belief, particularly in light of Heisenberg's uncertainty principle, which explains that the observer intrudes into the world of the observed and hence influences it.

On a more personal level, our experience in relationships should inform us that how we see other people is undoubtedly influenced by how we feel about them and interact with them. We don't necessarily view our partner, for example, in an unchanging objective state, but rather through a dance of feelings and emotions informed by our partner's and our own present and past. The positive connotations of objectivity that I mentioned at the beginning of this section—fairness, absence of bias—are relatively meaningless in this context.

HOW THE MYTH OF OBJECTIVITY UNDERMINES PSYCHOTHERAPY

The myth of objectivity underwrites the pathologizing of people throughout much of society. (I should explain that *pathologize* is a clinical term meaning to view or characterize certain behavior or characteristics as medically or psychologically abnormal, as when Dr. Rebecca Orleane writes, "Women's natural hormonal shifts have been pathologized into a diagnosis of hormonal disorder."[2]) Informed by the mechanistic perspective, classically trained therapists believe that with sufficient information they can drill down to the root causes of an individual's afflictions and render the appropriate diagnosis. As a result, the therapeutic gaze is often focused on the purported cause of a client's problems, which has led to reductive thinking—derived from Descartes's teaching.

For the most part, traditional psychology, including psychotherapy, rests on the foundation of diagnosis, informed by the biomedical approach. Indeed, health insurance companies require a specific diagnosis to justify coverage. The basic tenet of psychological diagnosis is that objectivity exists and that every psychological condition can be assigned a number that appears in a diagnostic guide produced by the American Psychiatric Association known as the *Diagnostic and Statistical Manual of Mental Disorders*, or DSM.

To take one example, the latest edition, DSM-5 (fifth edition, released in 2013), has added prolonged bereavement as a diagnosable mental illness. (The technical term is *persistent complex bereavement disorder*, or PCBD.[3]) The psychiatric "powers that be" have determined that grieving should conform to a prescribed period of time. The heartbreaking loss one feels over the death of a loved one is not considered a normal human experience, but subject to a formula that needs to fit into diagnostic and pathological criteria.

In order to diagnose psychological conditions, we have to assume that clinicians' subjective interpretations aren't getting in the way and that objectivity actually prevails, allowing the clinicians to inform the insurance company precisely what is wrong with each individual. From this perspective, a dozen clinicians working with the same individual would all render the same diagnosis. I can assure you that no such result would occur. Each of us sees through the subjective filter of our own life experience, colored by our beliefs, thoughts, personal history, prejudices, biases, and unconscious stirrings. This is as true for therapists as it is for their clients. For the most part, therapists are not calculating and detached automatons but educated professionals doing their best, yet constrained by an outmoded model of thinking. To that end, the field of psychology has not kept pace with the remarkable advances in the emerging sciences and, like the general public, clings to the principles of the classical paradigm. As Jungian analyst Marie-Louise von Franz wrote, "A psychology that does not keep up with the advances made in other sciences seems to me to be of little value."[4]

Steeped in the diagnostic penchant, traditionally minded therapists are confined to identifying and then treating the diagnosis or pathology, rather than cocreating new possibilities with their clients. Working from this methodology, they often cannot help move their clients forward—to get from *here* to *there*, or even to identify what *there* might look like.

People often have a reasonable ability to understand how they've become who they are and to appreciate the nature of their struggles. A continued replay of these life events, without sufficient focus on relief, if not transformation, leaves many individuals dissatisfied with their therapeutic experience. People who come to see me after having previously been in therapy often share their hope that they won't, once again, have to recount their life experiences. What they need is to see their lives through a new filter, one brimming with new possibilities. This requires a therapeutic approach no longer mired in reducing individuals to the wounds of their past, a perpetual reiteration of their confining wave collapses. New possibilities can be found in a humanistic venture of identifying and actualizing defining moments that usher in the clients' greater potential, relieving them from their limiting beliefs.

AN EPIDEMIC OF SUFFERING

According to the Centers for Disease Control and Prevention (CDC), twenty-five percent of the US population will suffer from mental illness in any given year, and that statistic grows to fifty percent over one's lifetime.[5] The enormity of that number is staggering and suggests that we've acclimated to a new norm—one of mass disquiet caused by our emotional and psychological suffering. Yet it's rare to hear anyone ask why this epidemic is occurring. Rather than asking how to best treat emotional and psychological problems, we should first be inquiring why they are so prevalent. The former question sets up an entire industry for treating the symptoms, while the latter has us look beneath the surface to discover the source and the possible solution.

If half of our population fell gravely ill, and we didn't understand the cause, we'd be searching for the answer with great urgency. The CDC would be working overtime as they did combating the Ebola virus. Shouldn't we be doing the same with our national emotional and psychological crisis? Unfortunately, because we've been acclimatizing to it for so long, we've become somewhat anesthetized to a phenomenon that should be causing us alarm. When we adapt to conditions that should be unacceptable and then normalize them, it suggests a deeper problem. We've become blinded to our own suffering. This phenomenon of making normal what should be seen as abnormal is known as *normosis*. The French writer who coined this word defined it as "pathology of normality, characterized by conformity and adaptation, in large scale, to a morbid context."[6]

This epidemic of emotional and psychological unrest has at least two main causes. First, a deep despair has been created by living somewhat incoherent lives as strangers in a strange land: we are living by the mechanistic blueprint that was never designed for accommodating human existence. We then exacerbate the suffering by labeling those who do struggle—arguably the majority of the population—as having something clinically wrong, and we diagnose them with disorders. The whole concept of "dis-order" comes from drinking at the fountain of separation and certainty. We might do better to look at disorder as if it were a fever fighting an infection. In this case, the infection should be seen as a worn-out worldview that no longer serves us.

As we've seen, the philosophy of Descartes set in motion our dependence on analysis and measurement, which taught us that by compiling enough data we could predict the future and, by extension, control and master our lives and our environment. Recall that Descartes's approach trained our thoughts to keep dividing things up so as to know all of the smaller parts. These teachings entrained our thoughts to simplify, reduce, and analyze at the cost of apprehending the bigger picture. Given this overreliance on analytical thinking, the epidemic of anxiety was inevitable since our minds are never at rest.

Analyzing should be a device in the mind's toolbox, but when it's the only one we reach for, we suffer the consequences. This indoctrination has resulted in excessive worrying, with the ensuing loss of presence that is essential to a balanced, harmonious life. Anxiety speaks to our relationship with fear. If your thoughts perpetually seek out your fears, you will naturally suffer from anxiety. You may recall the story of Tom in chapter 2; his need to assure himself of all outcomes, both at work and in his marriage, led him to suffer from chronic anxiety. The addiction to thoughts of this nature would be akin to walking through a thunderstorm carrying a lightning rod.

Those who suffer with anxiety often become engulfed in their thoughts' relentlessness, which imprisons them with a flood of despair. If you see yourself in this description, please recall the exercises from chapter 5 that explain how to see your thought and not *become* your thought.

Living under the mechanistic tenet of separation isolates us from one another and the universe at large. And removing ourselves from the inseparable flow of wholeness is utterly damaging. It perturbs our relationships as we experience life through the lens of false self-interest. (I use the word *false* because our sense of self is buoyed by our connectivity with others.) Separation reduces compassion and empathy and deprives us of the joy that comes from relatedness. Any sense of purpose and meaning becomes obscured through the filter of separation. The result is often depression.

DIAGNOSIS DISORDER: SUBJECTIVE THOUGHT MISTAKEN FOR OBJECTIVE REALITY

The second major cause of the epidemic of psychological and emotional unrest is the medical and psychotherapeutic drive to attach a label to what might otherwise be normal human behavior as we confront typical life challenges. As we noticed in the new DSM diagnosis for bereavement, the compulsion to affix a diagnosis to normal conditions has run amok.

The penchant for diagnosing what are ordinary responses to difficult life challenges as clinical depression thwarts our ability to help those going through those challenges. In many instances, depression is personally reasonable. Loss of a loved one, a serious illness, or being fired from your job creates painful circumstances. A woman in her thirties, named Lucinda, came to see me, saying that she was suffering from depression. She revealed that her husband had left her for another woman; she was now responsible for three young children, but not having yet secured a settlement agreement, she couldn't pay her bills. Her prior therapist had diagnosed her as suffering from depression and put her on antidepressants.

She did feel depressed and anxious. If she hadn't, I might have wondered if she were being avoidant. After all, Lucinda was suffering from heartbreak, abandonment, financial crisis, and parental overwhelm. But her depression was situational, not clinical. As far as I could deduce, she did not have a disorder. So I did not try to treat her depression, which, again, seemed appropriate to her circumstances. Instead, I assisted her in navigating her challenging circumstances, processing her feelings of hurt and abandonment, envisioning a new life, and gaining the tools to achieve it.

I'm not suggesting that there aren't people who are indeed suffering on a clinical level, but rather that the indiscriminate manner in which diagnoses are meted out without proper judgment and context is absurd. When diagnoses are delivered in the astronomical numbers we witness in America, it speaks to something much larger: a diagnosis disorder.

I would like to propose a new disorder for the American Psychiatric Association to consider in its DSM: confusing the diagnosis with being a real thing unto itself. The problem with all diagnoses (or any other type of label) is that when we confuse the description with being an actual entity, we exacerbate the problem. Once again we see literal thought operating. Given the illusion of objectivity, a psychiatric diagnosis should be descriptive, rather than pretending to depict an immutable reality that is observable by everyone in all contexts.

Taking an abstract idea—a diagnosis, for example—and turning it into an actual thing is known as *reification*. The philosopher Alfred North Whitehead coined the term "fallacy of misplaced concreteness" to describe the same tendency.[7] Diagnoses are literal thought claiming objective truth about something, all the while denying that thought created the construct it is now affirming. This is somewhat akin to the concept, derived from quantum physics, of observation summoning forth the reality. Recall that prior to making the observation, all that exists is a state of potential. The act of making a diagnosis constructs its reality. In this way, diagnoses take on a life of their own. Mind, through the collaborative efforts of a few psychiatrists, creates something—such as the term *depression*—and then denies its own participation in having done so.

Here is yet another example: In 1987, editors of the DSM-III-R (third edition, revised) *invented* the term *attention-deficit hyperactivity disorder* (ADHD) to describe a certain category of behavior that appeared to be presenting in the general population. This term was intended to describe certain behavioral patterns that professionals might detect in people, to aid in their treatment. Prior editions of the DSM, dating from 1967 to 1980, had referred to presumably similar "disorders" as "hyperkinetic reaction of childhood" and "attention-deficit disorder (ADD) with or without hyperactivity"; the diagnoses in later editions imply that earlier diagnoses were insufficient. Our thought is accountable for having created the term ADHD and for the fact that our subjective perception of what we see corresponds with the description itself. Again, we see literal mind operating as it creates the existence of something, changes it over time, and then denies that it constructed it in the first place.

If I hear a colleague say, "Emily has ADHD," I may respond, "How can someone have something that doesn't exist? You mean that you see something in Emily's behavior consistent with what we call ADHD." Notice the switch from literal thought to participatory thinking; the latter is the way we might articulate diagnoses from the participatory worldview. From this vantage, thought sees its role and notes subjective perception rather than deluding itself with objective statements. Diagnosis should describe what we think we see, nothing more.

In the former example, Emily appears to have an affliction, yet it's not as clearly discernible as pancreatic cancer, high blood pressure, or the West Nile virus. Her symptoms of what we call ADHD may be informed by her diet, her penchant for multitasking, the chaotic environment in her home, and an innumerable list of other possibilities. The diagnosis is a matter of subjective interpretation and needs to be acknowledged as such. If it's not, clinicians may fall prey to seeing this disorder wherever they look for it and may become further biased in their diagnoses. This conflating of conditions to have them correlate with the diagnosis is pernicious, as it doesn't allow for context and cultural differences. Taking the complexity of human behavior and reducing it to a diagnosis is simpleminded. This tendency is illuminated by Ethan Watters in his book *Crazy Like Us: The Globalization of the American Psyche*, in which he describes how DSM diagnoses are applied to diverse and indigenous cultures even though they don't present disorders that comply with Western culture. "We should worry about the loss of diversity in the world's differing conceptions of treatments for mental illness," he writes, "in the same way we worry about the loss of biodiversity in nature."[8]

A supposedly objective diagnosis speaks to the symptoms but rarely to the various contextual influences that inform the diagnosis. I was working with an individual named Enrique who had self-diagnosed—correctly, according to DSM criteria—that he was suffering from ADHD. He was indeed having difficulty maintaining attention and was easily distracted, but as I searched more deeply, it occurred to me that his self-esteem issues might be contributing to his ailment. Ordinarily, clinicians are quick to suggest that ADHD negatively affects one's self-esteem. But could that also occur in the converse? Could low self-esteem result in the diagnosis of ADHD? When Enrique had to read a complex report at work, he often felt challenged by his reading comprehension. His thoughts would then begin a critical monologue. "I can't believe I'm so stupid that I don't understand this report. I'm sure everyone else gets it."

Enrique was replicating one of the confining wave collapses from his childhood regarding his intelligence. He recalled being ridiculed by classmates and called stupid when he had difficulty reciting a passage from a book. Over time, he learned to avoid the disconcerting feeling that came from reading his work material by distracting his attention and leaving his desk. This distractedness provided him temporary relief from his negative self-talk. His attention deficit was merely a compensation that helped him avoid his self-esteem challenge. To diagnose him with and treat him for ADHD would be to miss the mark widely.

Low self-esteem doesn't appear as a diagnosis in the DSM, yet I find it at the root of most so-called disorders. Similarly, I've worked with many individuals over the years who have suffered from distractibility, which, upon deeper consideration, was the result of avoiding tasks or conversations that made them feel anxious. One individual in particular had complained that he couldn't remain present in social settings while having conversations with others. When he began to work with me, he informed me that he was on medication to treat his ADHD. As I delved deeper into his circumstances, I learned that he had significant social anxiety and had split part of his attention away from the conversation with others to monitor his own words in a self-critical way. In the midst of a conversation, his thoughts would wander off, and he'd think, "I can't believe I just said that. I wonder what they're thinking of me." He did indeed have attention deficit, but his primary challenge was anxiety, and once again self-esteem presented as the pivotal obstruction to his proper attention.

Many therapists focus on more specific diagnosable illnesses rather than marginal self-worth because we have medications we can prescribe or recommend for the diagnoses, notwithstanding their questionable results. But there is no pill to offer someone with low self-worth, so no revenue can be generated.

When people are feeling depressed, anxious, and lacking in energy, they need to address what informs these feelings and not simply suppress them. If we see feelings as a signal that something is off, we might use the feeling as a path to catalyzing positive change. In my practice, rather than treat the so-called symptoms, I prefer to assist people in coming to terms with their life challenges. It is essential to treat the person, not the diagnosis, and at all costs refrain from reducing anyone to a clinical compilation of symptoms.

WHY DIAGNOSTIC THERAPY CAN FURTHER THE DAMAGE

I was working with a man in his twenties named Timothy, who shared that he had been severely depressed his entire life. He was isolated, borderline suicidal, and had dropped out of college. As we reviewed the history of his therapeutic experiences, he recounted a detached and mechanical series of interventions during which all his psychologists and psychiatrists diagnosed and medicated him, suggesting that the best they could do was to help him manage his depression. By doing so, they contributed to his belief that he was a damaged human being.

First, I sought to understand on an empathic level what this word—*depression*—felt like for him. I wasn't interested in the clinical term but what the word signified in his mind. He recalled a traumatic period in his childhood when his alcoholic father was emotionally and verbally abusive. This confining wave collapse jolted him, so he adopted a coping mechanism that let him compensate for feeling at risk. In order to feel safe, he tried to ensconce himself in a womb-like protection by disengaging from life and relationships on any intimate level. He began to live like a recluse, avoiding interpersonal contact whenever possible. He didn't enjoy friends, hobbies, or sports and continued to isolate himself into young adulthood.

I suggested that in his circumstances feeling depressed made perfect sense. Why wouldn't he be depressed? He was barely living. His depression was merely symptomatic of his lifeless experience. Given his previous therapeutic exposure, I presented him with a radically new belief: that he might potentially be able to overcome his affliction. I suggested to him that he wasn't damaged as much as he was damaging himself. The power to access a defining moment, with all its incumbent possibilities, was altogether possible once he could shift his perspective and realize that he was inflicting isolation on himself and enduring its consequences. I was trying to have him buy into his possibilities.

His recurring thought that "there's something wrong with me" further affirmed by his therapeutic experiences, had become a self-fulfilling prophecy. Something was indeed wrong with him: his primary beliefs and ensuing thoughts denied him a reasonable life. Fortunately, he was open to reconsider his core beliefs as I illuminated how his literal thoughts—"I am a damaged person"—locked him in. In particular, Timothy grasped the value of participatory thought, and in the midst of a session one day he said, "I keep having the same thought that tells me how damaged I am." From there, he could appreciate that he had been damaging himself. Timothy is now employed, getting straight As in college, and enjoying friendships. He accomplished this significant gain in his life by breaking past the "objective truth" about himself.

Once again, the drive toward so-called objectivity obscures our fundamental participatory role in reality making. My former client Sam was fond of saying, "I have always been clinically depressed." His beliefs and ensuing thoughts and feelings continued to cast him as commensurate with his diagnosis. I eventually helped him to reframe his belief and say instead, "I've always seen myself as depressed, and I guess that's why I've felt depressed." The vital difference is that once he could see his thoughts' role in his experience, he opened the door to a new possibility. The Possibility Principle in such cases can be framed like this: I am not the equivalent of my diagnosis. The diagnosis is simply a set of words used by someone to describe what they think they see about me. I am free to grow, to evolve, to reframe how I think and experience myself. I need not be reduced to the label that has been affixed to me.

When individuals present an emotional or psychological issue, therapists typically look into their personal biography. They ask what the client's childhood experiences were like: were they sufficiently nurtured by their parents or were their caregivers neglectful, cold, or abusive? Therapists seek to put their clients' struggles into a historical context, and we should do the same in regard to how our culture is informed by its underlying worldview.

The world is not a machine but a living, evolving organism. The shift of mind from mechanistic to participatory worldview provides the pathway for our healing. With this shift, anxiety, depression, and our other afflictions retreat to a marginal percentage of the population, and the epidemic retreats. As the philosopher Thomas Kuhn asserts in his landmark book *The Structure of Scientific Revolutions*, the crises that appear within a paradigm are often resolved by a paradigm shift. If that's the case, we need to hasten our shift.[9]

8

FROM BEING TO BECOMING
Creating Authentic Self-Esteem

From the perspective of Newton's world of objects, everything is composed of inert and static things. And when we accept the Newtonian perspective, we see everything as, well, *things*, ourselves included. What an indignity to our human nature! Yet as we have seen, the emerging participatory paradigm depicts a universe in which flow is the norm, and all parts of the universe are evolving dynamically and inseparably. No part is left out. Reality presents itself as process as opposed to separate objects. This is wholeness unfolding, an ongoing state in which everything and everyone participates. In a participatory universe, the creative potential is the new mantra that supplants inert "thingness." Wholeness unfolding evokes a spirit of perpetual movement, refreshingly uncertain and replete with potential and inspiration. When thingness is replaced by flow, static states of being become an illusion; indeed, our state of *being* gives way to the process of *becoming*.

To ask "Becoming what?" is to seek a *thing* type of answer. What are the waves in the ocean becoming? What we see as material is only the temporary consolidation of energy, destined to return again to energy at some future time.

For all conscious entities, the process of becoming puts us squarely in this new paradigm of personal evolution and participatory change.

Some people eagerly seek the transformative process, while others ask why they have to change. This difference in outlook depicts their worldview: being or becoming.

FREEING OURSELVES FROM "WHO AM I?"

The fear of change roots many of us to fixed attachment to our identity. This anxiety often prompts people to ask the age-old question, "Who am I?" This very question suggests that there might be a plausible answer, as if our identity could be reduced to a fixed description. Individuals who ask this type of question are typically struggling for a core sense of self and are grasping to find a concrete answer. The paradox is that the more you seek to solidify who you are, the more fragile you feel and become. We would be better served to ask ourselves, "How would I like to experience my life?" The former question focuses on a fixed state (being) and the latter on process and flow (becoming). Recall the dilemma that Heisenberg discovered in regard to determining movement versus location. If we focus on the location (who am I?), we must lose the movement (how would I like to experience my life?).

We need to shift from concepts of fixed identity to a vision of process. Rather than taking a frozen snapshot of our identity, we should embrace an unfolding sense of self that has us perpetually reframing, recrafting, and rethinking ourselves and our experiences. The process of becoming enables us to move beyond our confinement from previous wave collapses and into our defining moments. As you've seen in many of my client narratives, when these individuals broke free of their past and ventured into new terrain, they accessed new possibilities. The process of *becoming* lies at the heart of the Possibility Principle.

As we seek to know ourselves, in all of our complexity, we should also devote our attention to the unfolding process of life. We might well consider how both our past and our interpretation of it have informed our present. Rethinking our past and placing it in a new context allows us to craft a different present and future.

Often a sense of insecurity is what has us inquire, "Who am I?" Imagine that you've been imprisoned for twenty years, incarcerated since the age of eighteen. You literally have no adult life experience outside of the penitentiary, and so your sense of self is extremely limited. You are about to be released from prison. The question "Who am I?" would provoke a fragile sense of self that might leave you apprehensive about your impending freedom. Yet it's unlikely that you'd choose to remain behind bars until you could secure your identity. You'd have little choice but to move forward into the uncertainty of postpenitentiary life and welcome your experience of becoming. The process of becoming necessitates your learning to get out of your own way so that you can embrace your natural unfolding.

At the other end of the identity continuum are those who claim to know themselves well and to have a clear sense of their identity. This group of people also manifests a deep fragility about their identity. To know yourself so well leaves little room for growth. It speaks to a fixed sense of self. Even more, it suggests a protective mechanism that guards against deeper reflection and embracing change. If I'm dead certain I know exactly who I am, then I see myself as a fixed entity, stuck and inert.

It's wise to self-reflect and invite introspection, but doing so requires maintaining a fine balance. Be cautious not to fall prey to overanalyzing. The goal is to maintain malleability as you engage in reflection, as if you were a willow tree rather than an oak tree. The willow is flexible and survives the storm as it bends with the vicissitudes of its environs, whereas the rigid oak is more likely to crack open. When you maintain flexibility during reflection, your reflection is more contemplative and forward-looking, allowing you to unfetter yourself from the imprint of old injurious wave collapses. Try to envision how you'd like to experience your life and note the aspects of yourself that you'll need to let go of. Then look at the core beliefs and recurring thoughts that keep reinforcing your limitations. Work with that dissonance as you release your past.

Recall that embracing uncertainty enables us to join in with the universe's perpetual flow of energy. The process of becoming is forgiving. In the flow of becoming, we are no longer rooted in the hardship of fear, insecurity, or concerns about mistakes. The fuller participation in our unfolding life assists us in the art of living well. Becoming is open and infinite; being is structured and limiting. We must learn to live in a manner that permits us to see our personal evolution as a process in which we must be open to inquiry and learning and always receptive to new meaning. To more deeply engage the process of becoming, we need to shed the tired parts of our personality that may no longer serve us. These archaic aspects of ourselves prevent us from spiraling up into new processes of becoming.

OVERCOMPENSATION PRODUCES IMBALANCE

Several years ago I broke my foot as I missed a step on my front porch. The break occurred on the outside part of my foot, the fifth metatarsal, and my doctor told me I wouldn't need a cast. In deference to the pain on the outer perimeter of my foot, I shifted my weight toward the inner perimeter. By the following week, I had overstressed the unbroken part of my foot by placing an inordinate amount of pressure on it. I actually experienced more acute pain in that area than in the broken area. A month later the broken bone had essentially healed, but the damage I caused to the inner part of my foot still lingered. I had overcompensated for my injury, leading to a debilitating imbalance.

This same tendency to overcompensate also provokes havoc in our emotional and psychological lives. At different times in life—particularly in childhood—we develop mechanisms to cope with the challenges, wounds, disappointments, and hardships we encounter. Coping mechanisms are unconscious but manufactured adjustments that we make to our personalities as we try to shore up those parts of us that feel insecure, insufficient, or threatened. Our confining wave collapses set in motion these self-protective devices, or personality masks, in response to disturbances. We're not typically aware that we're constructing these personality masks because they assimilate into our being in subtle ways.

An abusive, neglectful, or unloving parent's hurtful actions may prompt us to react indifferently or callously so that we can survive the pain. Or we may swing to the other end of the spectrum, compensating by taking on the role of caretaker to ward off the damage or to extract some of the nurturing that is missing. A chaotic or turbulent home environment may induce us to fashion the mask of being a people pleaser, trying to placate everyone so that peace may reign. In extreme cases, we may become overly dependent in our relationships, seeking what we didn't have in our childhood. Embarrassing or traumatic events that make us feel humiliated may also catalyze these personality adjustments.

I had been working with Alex for a brief time when he revealed that during his childhood he experienced his father as extremely volatile. He soon learned to be hypervigilant so as not to trigger his father's anger. When something troubled him about his dad, he suppressed it. This coping mechanism remained an unconscious part of his being, and Alex had no idea how deeply it affected his marriage decades later. He came to feel that his wife was emotionally distant and harshly critical of him, but instead of sharing his upset with her, he sublimated his hurt feelings and picked on her over marginal issues. This constant nit-picking caused her to distance herself even further from him. His compensatory behavior, instigated by his childhood coping mechanism, was sabotaging his marriage. Even worse, because he continued to suppress his real issues with his wife, they had no opportunity to resolve them. What had kept him safe in his younger years was now setting up a dysfunction in his primary relationship. His coping mechanism caused a serious imbalance. I see this phenomenon in virtually every couple I work with: the couple's conflicts grow out of each individual's unresolved challenges that spill onto the relationship.

Our coping techniques are adaptive and may be advantageous when we first adopt them. They get us through the storm, so to speak. The problem is that they become hardened and inveterate over time; they imprison us as they block the emergence of our authentic self. The past leaks into the present, as it did for Alex, and what was once a coping mechanism becomes a suit of armor. As we clank through life wearing this outdated armor, it thwarts our process of becoming.

Psychological coping mechanisms are a means of compensating for a deficiency, the same way I tried to compensate for the broken part of my foot by shifting my weight to another part of the foot. But when we overcompensate, physically or emotionally, we become imbalanced. For example, I've seen many doting, excessively attentive parents burden their children as they overcompensate for the lack of proper attention they received in their own childhood. The pendulum swings too far.

Coming into balance—the ultimate goal in achieving a life well lived—requires noticing where we are overcompensating and then freeing ourselves from the masks and coping mechanisms we developed, because of our confining wave collapses, so that we can embrace our vulnerable self, which has become obscured in our zeal to protect it.

If you feel you may be overcompensating in some way, first reflect on what vulnerable part of yourself you're protecting. To illuminate these fragile aspects of yourself, ask, "What part of me feels insecure or deficient in some way?" For example, you may feel overly sensitive about others' opinions of you, or you may feel unsafe confronting issues with other people. Or you may feel that you're not smart enough.

Once you've identified the vulnerable part of your being, ask yourself what your overcompensation looks like. If you are sensitive to others' opinions of you, you may have told yourself that you prefer being a private person. If you feel unsafe in confronting others, your personality may have morphed into being a people pleaser, allowing you to avoid potential discomfort. And if you feel you're not smart enough, you may compensate by not asking questions out of fear of revealing your ignorance.

The answers to these questions will clarify where you are imbalanced and will signify the confining wave collapses that you may be overcompensating for.

Once we have identified the tired parts of our personality that have outlived their function, the old coping mechanisms that no longer serve us, we can gain access to the potential that a new and defining wave collapse can provide; we can come back into balance and engage in the natural process of becoming. Yet, as we saw in chapter 5, the uncertainty of the new terrain often evokes discomfort. One of the primary reasons for that discomfort is our attachment to our identity. Although we clearly see the obstacle to our growth, the prospect of losing an old yet tired feature of our identity may feel daunting and can even provoke anxiety. At the root of our discomfort is the often-unspoken question, "If I'm not who I think I am, then who would I be?" which is simply an extension of the question, "Who am I?"

THE STRENGTH OF VULNERABILITY

The scourge of our times is low self-esteem. An enormous percentage of people have come to believe limiting and negative stories about themselves, and they experience their lives accordingly. The more they do so, the more they try to hide, disguise, or overcompensate for their insecurity. The problem is that their coping mechanisms keep them stuck and impede them from joining in the process of becoming.

Exacerbating this problem is our misinformed cultural meta-narrative that demands the appearance of strength. We are taught to act strong and to hide our vulnerable side. This tendency results in large part from the individualistic and competitive instincts derived from Newton's theme of separation. Being separate sets up measuring and comparison.

This messaging promotes fear and exacerbates our insecurities as we hide our inner self from others, lest they judge us; in turn, we abandon ourselves and defer to others, decimating our sense of self-worth. When we act or pretend to be different from what we truly are, we abandon our real self and put on a mask in an attempt to control what we think others will think of us. We manipulate and camouflage our self as we seek the approval of others, or at least try to avoid their disapproval. This sets up our primary betrayal of our genuine self.

The pathway toward self-value requires embracing our vulnerability. Anyone who is comfortable with their vulnerability has nothing to hide from others and is genuinely strong. People who don't occasionally struggle with self-doubt, fear, or insecurity are a small minority. This is a normal human experience, and we should engage it as such, without embarrassment or apprehension. Embracing our vulnerability means acknowledging our fears and insecurities without judging them. When we do that, we have nothing to hide, and we liberate ourselves from setting up others as our judges. We must embrace our vulnerability to attain inner strength.

For me, the word *vulnerable* doesn't elicit weakness or fragility, but openness. Hiding our true self from others makes us fragile. Being our true self makes us strong.

This myth that vulnerability is a weakness, not a strength, not only limits our lives, but also disconnects us from one another. When we decide to hide our feelings from one another, we live out our lives falsely, thinking that our shortcomings or self-doubts are unique to us. Yet those same individuals whose opinions we are so worried about are probably doing the same thing. The vast majority of people are disempowering themselves, thinking that others are more confident and secure than they are. The tendency to hide from others becomes apparent when I'm facilitating group work on any number of topics. When one individual begins to unburden herself and share her issues, I see from the expressions on the faces of others that they are connecting empathically. The dominos start to fall as, one by one, others also share what they had been locking up.

RETHINKING SELF-ESTEEM

I believe that the term *self-esteem* is often misapplied. The first half of the expression, *self*, indicates that *esteem* is derived from one's inner being. Yet most people seek a sense of worthiness from things that lie outside of themselves. For a student, it might come from good grades; for a businessperson or worker, it may derive from a promotion or a raise; and for most individuals, praise or acknowledgment provide a temporary increase in esteem. Our society generates billions of dollars in revenues from inducing people to seek the quick fix of vanity as a means toward feeling better about themselves. Yet none of these affectations of glamour or success actually contributes one iota to real self-esteem. Paradoxically, they may undermine our self-respect.

If we contort our personality to seek recognition or approval from others, or to avoid disapproval, we're pursuing what I call "other-esteem." We're trying to feel better about ourselves by being disingenuous. The more we do this, the further we move from genuine self-esteem. Being approved of or valued by others may make us feel good, but if we betray our authentic self in order to manipulate these results, we decimate genuine self-worth. Not only that, but when we act in this manner, we are taking our well-being and serving it up to other people. It then becomes the other person's duty to decide if we are worthy. This is not a healthy place to be, and it is a soul-defeating exercise.

The simple truth is that others can't judge you. People have opinions of you; that is entirely natural. To elevate their opinion to the status of a judgment, however, is counterproductive. No one can judge you unless you confer upon him or her the power of being your judge. The only person who arbitrarily has such power presides in a courtroom and wears a long black cloak; all others are people with opinions. And with a healthier measure of self-esteem, we might more easily tolerate others' opinions without elevating their beliefs into construed judgments and objective truths.

If someone else doesn't grant us their approval, we have a habit of claiming that they rejected us. In truth, we have rejected ourselves when we set the other person up as judge. You might recall how this happened with Helen, the client I introduced in chapter 3, when she set up my colleague Jim to be her judge. The degree to which we are reactive to others' opinions of us is inversely correlated to our level of self-esteem.

Most parents would claim that they are thoroughly invested in their children's self-esteem. Educators, therapists, and guidance counselors also place great value on the development of children's self-worth. Yet

I would argue that most don't begin to comprehend self-esteem. If an A student becomes depressed by getting a B, it is abundantly clear that the student's esteem is contingent on their performance. Performance should be seen as the icing on the cake, but the cake, so to speak, is the student's relationship with him- or herself.

Rethinking what genuine self-esteem looks like allows us to live more sensibly. Authentic self-esteem is the legitimate foundation for a healthy relationship with others and with ourselves. It removes the construct of neediness so prevalent in most relationship challenges and liberates us to thrive as issues of rejection and judgment recede. Most of what you've learned so far in this book is designed to help you live a fearless and resilient life—cornerstones for self-esteem. The fundamental building block to authentic self-esteem is facilitating your full engagement in the process of becoming. And as we've seen, breaking free from old coping mechanisms and opening to defining moments are the first step.

When our sense of worth emanates from within, we engage in our greater authenticity by inviting our vulnerability—the openness that enables us to join the flow of life.

RETHINKING MISTAKES AND FAILURE

Allowing our identity to evolve and change as we break free of old, worn-out encumbrances often induces anxiety about, if not fear of, making a mistake. This anxiety is rooted in the old paradigm of *being* as opposed to *becoming*. When we see reality operating from a fixed place, we unthinkingly try to hold on to what is and ward off uncertain change. This is the equivalent of trying to hold back the flow of life. From the vantage of being, the more we seek certainty and predictability, the greater is our fear of making mistakes. From the vantage of becoming, we can naturally loosen our grip around this anxiety, as we appreciate that the consequences of our decisions may often not be as fixed as we imagine them to be. Freeing ourselves of the fear of mistakes, paradoxically, usually enables us to make clearer and more sensible decisions.

Our fixation on certainty and predictability keeps us defensively resistant to change and fearful of this construct that we call a mistake. Generally, what we call a mistake is a decision or an action—or lack of one—that we regret. Mistakes usually cause some degree of pain, loss, or struggle. We don't like the consequences, and so we call it a mistake. The paradox is that the consequences that we try so hard to avoid may be precisely what we need to experience. *A mistake is simply an event, the full benefit of which we may not have come to realize.*

I've often heard people speak of their former marriages as "mistakes" because they terminated in a divorce. Yet without such a difficult experience, neither person would have had the opportunity to discover deeper truths about themselves and their choice of partners. When I reflected on my former marriage, I came to appreciate that my ex-spouse hadn't changed over the years we were married. This became a source of our conflict because I had changed quite a bit and expected her to do the same. Yet I had no reason to anticipate that she would change; she had never signed on to do so. Then why had I become discontent with her? This question prompted me to consider why I had asked her to marry me in the first place. I needed to go through these travails to reflect on aspects of myself that had remained beneath my surface. Marrying her wasn't a mistake; it allowed me to engage my struggle and come into a deeper consideration of who I had been as a younger man.

Insights born out of what we call mistakes are necessary for our psychological, emotional, and spiritual growth. They are also a fundamental part of our learning process. If we look at our lives as perpetually unfolding, a never-ending process of becoming, we might consider that what we call a mistake is simply a moment that we have frozen in time. But the flow of life doesn't freeze; it cascades on and on, whether or not our perspective enables us to see it that way.

The very construct of mistakes assaults our relationship with our self in that it deprives us of being at peace. When we think we've made a mistake, we tend to defer to the voices and opinions of others, just as when we devalue our self-esteem. In both cases, we suffer because we lose the creative, joyful spirit that more fully engages life.

The concept of a mistake is tied to the larger notion of failure. Just as there is really no such thing as a mistake, I would offer that there is really no such thing as failure. Imagine watching a toddler struggling to take his first steps, only to fall. How ridiculous would it be for us to proclaim that he failed! He simply hasn't yet mastered the skill of walking. Success has not yet been reached. It is altogether human to struggle as we move toward what we're trying to achieve. To refer to this process as failure is destructive and self-defeating. The notion of failure is simply a belief that the mind has created, based on the tendency of the classical worldview to measure and compare. The construct of failure is intrinsically rooted in the mechanistic paradigm. A machine may falter or fail as it ceases to function, yet we have once again conjured a mechanomorphism as we become machinelike. What we call mistakes or failures not only are learning opportunities, but also often open new doors for us.

I was working with Amy, a successful hedge fund manager who was nevertheless discontented in her life. She was adroit at her job, but she complained that it didn't feel like her true calling. She didn't feel joy or deep gratification from her work. One day she came in for a session fretting over an ill-advised investment position she had taken that cost her firm a considerable amount of money. In anyone's parlance it looked like a "mistake." This caused her a wealth of distress. It also opened the door for us to talk about what she felt was her true passion—art. Eventually, the fallout and discord from her investment choice opened the path for Amy to pursue her passion, and she opened what was to become a successful art gallery.

When we find ourselves terrorized by thoughts of mistakes or failure, we lose the opportunity to live more fully. No single correct decision or pathway exists in the process of becoming. Liberating ourselves from this fear enables our lives to unfold with greater purpose. Mistakes and failure are cultural constructs rooted in the either-or thinking that we've been exploring and that would have us believe that we must have simple and clear, right and wrong choices. Embracing a universe that is participatory, inseparable, and flowing allows us to transcend such limiting constructs. Just as importantly, it brings us into the power of new possibilities.

THE PROBLEM WITH PERFECTION

During my years as a therapist, I've treated increasing numbers of individuals who are driven to distraction through their pursuit of perfection. The desire to be perfect burdens people and imprisons them with unrelenting stress, often creating havoc in their lives, yet these people believe that seeking perfection is desirable.

Perfection suggests a state of flawlessness, an absence of defects. To be perfect implies a condition in which your actions attain a level of excellence that cannot be exceeded. Certainly a student can strive to attain a perfect grade. You can hope to bowl 300 or produce a perfect report at work. You would also certainly hope that your surgeon does a perfect job on your operation. Yet the goal of being perfect in life is altogether a different story. A machine or electronic device may operate perfectly—at least for a while. Over time, however, it will begin to wear down and require repair. In our desire to be perfect, we find another meme that is rooted in the paradigm of Newton's mechanistic universe. Humans were never intended to be perfect. That's part of the consideration of being human. Consider the expression, "I'm only human."

In counseling people who are beleaguered by their need to be perfect, I have grown to appreciate that their pursuit of perfection is really a disguise for their insecurity. The obsessive need for perfection is a way of saying "I'm not good enough just as I am." Most of these people received messages earlier in life—wave collapses—that they were lacking in some essential way. So they decided that only by being perfect would they be beyond reproach. With such an affliction, we might look at perfectionism as a compensation for these earlier wave collapses that corrupted our well-being and self-esteem. Perfectionists tend to think that other people are somehow better or superior to them, so they need to be without flaw just to catch up.

Instead of seeking perfection, we would be better off developing the ability to be fully present. Without any distracting thoughts measuring or grading us, we're free to be in the moment, when we are truly alive. Perfectionists typically aren't present because they're busy either critiquing the past and replaying their every decision or fretting about their future decisions. They might spend their time more effectively envisioning how to transcend the insecurity and wounds that catalyzed the desire for perfection in the first place. As we saw earlier, embracing your vulnerability, including all of your insecurities, is an essential first step in healing those wounds. It is also a sure sign that you are on your way from being to becoming.

The great American painter Albert Pinkham Ryder described his creative process with a fascinating image drawn from nature:

> Have you ever seen an inch worm crawl up a leaf or twig, and there clinging to the very end, revolve in the air, feeling for something to reach something? That's like me. I am trying to find something out there beyond the place on which I have a footing.[1]

Ryder's description also captures the emergent quality of becoming. Instead of striving for an illusory perfection, we should be striving to grow and stretch beyond our familiar zone and into the new possibilities that always await just outside of it.

THE WONDER OF "WHAT IF?"

We can catalyze our emergence from the narrow, programmed road of life and open ourselves to wonder and contemplation by asking a simple "What if?" As we proceed in our process of becoming, a sense of wonderment enables us to proactively engage our possibilities. Coming out of the prescriptive path rooted in determinism into the participatory role of recrafting ourselves fosters both authentic self-esteem and a mastery of our life experiences.

Upon graduating from college, I had a premonition that life might well resemble the conveyor belt that delivers our luggage at the airport. If you don't retrieve your baggage, it continues along its predetermined route. Similarly, once you step onto the conveyor belt of life, it just keeps going. Make a few decisions pertaining to career and perhaps marriage, and the rest of your life becomes somewhat directed for you.

By the time you recognize what's happening, half a lifetime may have passed, which is probably how we came to the term *midlife crisis*.

Ordinarily, not until midlife do we come to recognize that we haven't really been fully aware of the years flying by. The midlife crisis is simply a time-out allowing us to regain some directorship over our life. This often results in alarm from others, who may be trying to avoid such self-reflection and prefer to have us back on the path of the "predictable." With that thought in mind, I promised myself that every so often I would jump off the conveyor belt of my life and reevaluate how it was going. I'd contemplate what changes I might want to make. A pathway toward this goal would be to ask myself, "What if?" In truth, if it weren't for that contemplation, I'd still be condemned to my previous, unfulfilling career and certainly wouldn't be sitting here writing these words.

A few years ago my son, on graduating from college, was discussing his future plans with me. He said that he didn't feel ready to settle down to the routine of a nine-to-five job (more likely eight-to-six, these days) and was entirely uncertain as to what direction to take. He asked, "What if I could find a way to support myself while seeing more of the world for a while?" We discussed the possibilities and explored how he might accomplish that. His *what if* question manifested through his living in Prague and teaching English as a second language. His life unfolded from that experience; his future career path became evident through engaging in this adventure.

I should caution that those who are fearfully attached to the tenets of the mechanistic paradigm might experience the *what if* question differently. For them, this question often leads to an anxiety-based fear of unwanted outcomes. Once again, our confining and defining wave collapses and our worldview will inform which direction the *what if* leads us to: wonderment or worry.

For those with a quantum worldview, this question permits a remarkable unfolding. It suggests that things need not be fixed and deterministic but rather can be open to a creative shift. What if I changed my thinking or beliefs about this or that, or him or her? The very notion of *what if* opens up our creativity and enables us to live more masterfully. It's as though we reopen the possibility of our life becoming the canvas that it truly is—with the added bonus that we've put the paintbrush back into our hands. The absence of wonder and imagination dulls our being and compels us to live a regimented life, devoid of personal mastery. Asking *what if* opens the doorway to a more meaningful and participatory life. These words prompt our participation with our future, the full engagement of our experience of becoming. *What if* seeks new possibilities.

9

BEYOND THE MIND-BODY
CONNECTION

Those of a holistic mind-set often refer to the phenomenon of the "mind-body connection." Sometimes called the mind-body problem, this phrase suggests that we have discovered a link between our mental and physical being, and we need only to determine how mental states, such as beliefs and thinking, have an effect on physical states, events, and processes. Given my own holistic point of view, many people react in surprise when I challenge this concept and suggest that no mind-body connection exists.

Inseparability implies that divisions between mind and body are artificially drawn, if not outright incomprehensible. The only thing separating our skull from the rest of our body might be the tie that some men knot around their throats. If we proceed from the underlying assumption that nothing is truly separate from anything else, then the word *connection* loses its relevance. In short, no mind-body connection exists because the two have never been separate. In this chapter, I'll show how we have constructed a false reality in dividing mind and body and how the new science now provides the opportunity for us to restore the integrity of wholeness to humans.

With his axiom, "I think, therefore I am," René Descartes effectively separated mind and body, a division referred to as the Cartesian duality. This philosophy implies that mental activity exists in the brain and that the body is altogether cordoned off, resulting in a profound dichotomy with grave consequences.

Although the notion of a mind-body connection is a far cry from the dualistic approach of traditional medicine, it doesn't go quite far enough. In opposition to Descartes's message, I'm proposing—along with countless others—that mind and body appear to be merely differing aspects of the same whole (the human being). Just as the head and tail of a quarter are not separate but are different sides of the same coin, mind and body are thoroughly entwined and, as such, inextricable. If two photons, once entangled and then separated, continue to operate as a seamless whole, how could it be otherwise in the case of mind and body, which are not even separated?

Here we need to step back and consider how deeply brainwashed we've become by the paradigm of separation. Taking something that is whole and intact, we've amputated the head from the body and now believe that they are indeed distinct from each other, despite the obvious fact that neither one can function without the other in any practicable way (except perhaps in certain science-fiction movies). Descartes performed quite a magic trick to convince the world that this is actually plausible. Four centuries later we've continued to be a mesmerized audience that buys into his sleight of mind (pun intended).

Merriam-Webster defines *psychosomatic* as "of, relating to, involving, or concerned with bodily symptoms caused by mental or emotional disturbance." The word itself typically has a pejorative flavor, as if physical ailments conjured by the mind are less substantive than those with a purely corporeal cause. Yet the distinction that the term *psychosomatic* is making is unclear if not absurd. If something is in your mind, of course it's in your body as well! It couldn't be otherwise. Creating a word to designate a connection between parts of a whole entity illuminates the tragedy of duality. When we calm the mind by meditating, we note the effect it has on slowing our heart rate and reducing our blood pressure. Why would other circumstances of our mental activity be entirely cut off from our physical being? Why would we think that prolonged fear or anxiety wouldn't impact our physiology?

What we call the placebo effect is further proof that the mind affects the body. Although medical researchers at first marginalized the phenomenon by saying, "It's just a placebo effect," attitudes have changed. The popular Internet site WebMD now says this:

> The fact that the placebo effect is tied to expectations doesn't make it imaginary or fake. Some studies show that there are actual physical changes that occur with the placebo effect. For instance, some studies have documented an increase in the body's production of endorphins, one of the body's natural pain relievers.[1]

The prestigious Mayo Clinic shares that opinion, as published in its *Mayo Clinic Health Letter* in April 2014: "The placebo effect can vary from having zero to a 100 percent effect, even in the same condition."[2] Indeed, the placebo effect should motivate medical research to further study the mind's ability to heal the body. Our mind and emotions affect our physical being, just as our physical condition influences our thoughts and feelings. (Clearly illness, injury, and terminal disease can have a devastating impact on our state of mind.) The mental, emotional, physical, and spiritual planes all interpenetrate and overlap one another. An impact in one sphere has immeasurable effect on the others.

THE CONSEQUENCES OF SEPARATING INTELLECT AND INTUITION

From the worldview of an inseparable universe, the notion of connection, let alone separation, is merely a trick of the mind. The dichotomies that the mind constructs are caused by the binary nature of our thinking. Recall that mind fragments wholeness by creating divisions—mind and body, left brain and right brain, among others—and, having done so, compartmentalizes them so thoroughly as to wall them off from one another. The mind forgets that it actually made these compartments and sees them as real. Then, to our surprise, the differentiated parts nevertheless leak into each other.

We should be looking at the larger consideration: mind has constructed a false reality in dividing mind and body, and we live accordingly, causing ourselves much harm. The belief that our mind and body are separate silos begins to seem like mass hallucination. Moreover, these artificial divisions occur virtually everywhere we look.

One such artificial dichotomy involves intelligence and intuition. Intellect is valued as an expression of logical and rational inquiry. Most people believe that intelligence is a by-product of one's brain and utilizes the skills of logic and analysis. Intellectual acuity is the pathway of achievement and success—or so we're led to believe. One earmark of Western culture is that we grade and measure what we value most—test scores, income, economics, and almost every measure of traditional success—and we have even created measures of intelligence, including standardized tests and IQ evaluations. We have attributed the notion of masculine energy to intelligence rooted in the epistemology of classical thinking, philosophy, science, and gender archetypes (or stereotypes).

Intuition, by contrast, speaks of a way of knowing that is instantaneous and independent of rational cognition. We generally regard it as a feminine trait grounded in the emotional sphere. It is notable that we don't measure our intuitive capabilities as we do our intellect. The phenomenon of intuition is believed to have an immaterial quality, since it doesn't appear to emanate from a physiological organ—namely, the brain. For this reason, it is often devalued into something not appreciably more valid than a hunch. Intuition is generally considered to be a weaker sister to intelligence (except perhaps in the few remaining indigenous cultures). This isn't surprising given that the prevailing Western cultural paradigm still inclines toward the masculine.

Ordinarily, first-world cultures heavily favor the intellect over the intuition, most likely because of our emphasis on logic and our tendency toward analysis and reductive thinking, all components of the intellect. This prejudice reaches all the way back to Aristotle. But intellect, as understood via the thinking of Newton and Descartes, also severed wholeness and divided reality into separate and distinct compartments. So logic and analysis, the tools of the rational intellect, although essential in many circumstances, can wreak destruction once they've become the exclusive mode of making decisions.

These tools are insufficient for grappling with a wide array of nonmaterial subjects ranging from the nuances of relationship to matters of the heart and soul. This particular way of knowing is blinded to the unintended consequences of its actions and to its adherence to fragmentation. The intellect devised ingenious ways to extract fossil fuel from our planet and to create an incalculable number of chemicals to further our industrial growth, without stopping to consider their unintended consequences—including climate change and cancer. Intuition would probably be more sensitive to the relationship between our actions and their consequences, because intuition tends to consider consequences beyond the immediacy of simple cause and effect.

Intuitive ways of knowing are effortless—and they aren't subject to our mind's tricks. We have many ways of knowing, including cognitive, experiential, intuitive, spiritual, and transpersonal. Limiting ourselves to just one mode (intellect) impedes our ability to make considered judgments. Just as analyzing wreaks havoc if it's the only tool we reach for, intellect cordoned off from intuition can be equally insular and counterproductive.

An intellect that doesn't accommodate and integrate intuition is incomplete and, at times, dangerous because it fails to see its own participation in what it identifies as problems "out there." In the discussion of fragmentation in chapter 4, we looked at examples of this habit of fragmentation. Intellect operating without an intuitive capacity is insufficient and incomplete.

If intelligence, uninformed by intuition, falls well short of wisdom, then intuitive processes that deny intellect are also lacking a more thorough grounding and belie the vitality of intelligence. When we incline heavily toward one at the cost of the other, we limit our field of vision. Moreover, we prohibit ourselves from seeing and operating in wholeness because, operating from a fragmented perspective, we've once again chosen this or that.

REINTEGRATING INTELLECT AND INTUITION

Wholeness requires us to integrate the differing ways of knowing and once again bind the qualities that we've severed from each other. The rational tools of intellect have their rightful and necessary place in our intuitive process, but intellect is fuller and more expansive when integrated with intuition. Rather than seeing intelligence and intuition as opposing processes, we should see them as complementary qualities. I would call the marriage of the two *intellectual intuition*.

Although I'm aware that generalizing is a categorical pitfall, and that many exceptions exist for each generalization, I do find that men tend to root themselves in rational intellect while women are more inclined toward intuitive ways of knowing. At times, this dynamic becomes disruptive to relationships because the different genders appear to operate on differing systems of communication, which we'll examine in depth in the next chapter. When this occurs, both parties may experience an absence of shared meaning because the intellect and the intuition tend to battle with each other, creating an incoherent communication. It's as if the two individuals are speaking somewhat different languages. To that end, a balancing of masculine and feminine would be aided by each integrating the opposing aspect, as in the Taoist principle of yin and yang, which views opposing characteristics as complementary, such as light and dark, fire and water, heat and cold, expansion and contraction.

If you ask how to go about achieving this integration, I've found that balancing occurs most readily by simply setting your intention for it to occur. When we learn to quiet the fragmenting of our thoughts and to still our reactive mode, a space emerges that we weren't previously in touch with. Setting your antennae to summon your intuition, as it were, typically yields results. As overreliance on rational and analytical thinking retreat, the intuitive process gets a chance to percolate.

Most people orient toward an imbalance, identifying primarily with either thinking or feeling. Their thoughts or sentences generally begin with either "I think" or "I feel." People who are oriented toward feeling generally distrust their intelligence. They probably had confining wave collapses earlier in life that made them question their intelligence. By the same token, those of us inclined toward thinking probably grew up with the belief that feelings had little value or substance. An individual who integrates both thinking and feeling operates on a much more powerful and coherent level.

Ask yourself whether you more often use "I think" or "I feel." The answer will tell if you gravitate toward thinking or feeling, intellect or intuition. Once you identify the component that you're not in touch with, set your intention to integrate it into your fuller being. If you tend to favor thinking, start prompting yourself to inquire, "What am I feeling?" If you tend to favor feeling, ask yourself more often, "What am I thinking?"

I have been assimilating intellectual intuition into my work as a therapist for some time now. Indeed, this is increasingly the manner in which I see, work, and live. It provides a more expansive and immediate way of perceiving, whereas a reliance exclusively on either intellect or intuition would have me missing the mark far too often. When we learn to honor the wisdom of our intuitive processes and blend it into our intellectual capacity, we move from a one-sided perspective into a powerful integration of differing ways of knowing. The combining of intellect and intuition provides us with a somewhat lost art: wisdom. Wisdom is the ability to better perceive the unintended results of both our actions and inactions, enhanced by our intelligence.

Seeing in and through wholeness, facilitated by intellectual intuition, may at times lead to experiences that mirror the quantum experience of entanglement. This type of inseparability moves beyond the entanglement of a pair of photons or two individuals. It occurs when our mind—consciousness—and the material world are no longer separate but reconcile as one. In such circumstances, space and time merge and causality ceases to exist. The intuitive sixth sense opens a gateway to an essentially mystical experience of oneness, which correlates with what quantum physics has been reporting.

During the winter holiday season of 2004, I was vacationing in Baja California, Mexico, at a magnificent spot where the desert meets the ocean. These surroundings enabled me to reconnect with nature and tune in to the environment around me. As I did so, my sense of serenity heightened, and I became more at one with the physical world. This tranquility was, however, quickly tempered by reports of the Indian Ocean tsunami of December 26 and the humanitarian calamity that it wrought. Despite being on vacation, I felt compelled to read the daily reports of the disaster that was unfolding.

National Geographic reported that hardly any death to wildlife followed the wave. Indeed, surprisingly few animal bodies could be found. According to eyewitness accounts, "Elephants screamed and ran for higher ground. Dogs refused to go outdoors. Flamingos abandoned their low-lying breeding areas. Zoo animals rushed into their shelters and could not be enticed to come back out."[3] And yet, as many as 230,000 humans perished. The *National Geographic* article quotes one scientist who believes that "animals can sense impending danger by detecting subtle or abrupt shifts in the environment," and the author speculates that animals might, in fact, have a sixth sense—something that humans apparently were lacking.[4]

I left the beach and went to my room to begin writing an article about this notion. My thoughts went something like this: The sixth sense being attributed to animals must have previously been the natural state of our humanity. Animals had never been subjected to the fragmenting of wholeness perpetuated by mechanism. As an integral part of animals' ecological balance, their sixth sense never atrophied because of their reliance on fragmented thought born out of the paradigm of separation. Their way of knowing has remained essentially intuitive.

It makes perfect sense that wildlife anticipated the tsunami. The animal kingdom and natural phenomena are just differing aspects of the environment and, so, share an entangled state. They are as one. Humans, on the other hand, have severed our connection to nature as we've ruptured wholeness. By separating from and exploiting the environment, we've lost our relationship with it. We no longer feel a correspondence with nature—or the intuitive capacity to foresee what's coming.

SYNCHRONICITY

As I sat at my desk in Mexico writing my article on our forgotten sixth sense, a bird flew into my room through an open window and perched itself on the arm of my chair. For a few moments, which I experienced with a timeless awe, we simply took each other in. The bird then departed as unexpectedly as it had arrived. This was no simple coincidence, I was certain. It was synchronicity, a result of conscious material (my thoughts) becoming manifest through the physical world, or when psyche and matter become as one. Often, such experiences serve as a resounding affirmation that we are indeed on the correct path—in my case, as heralded by this extraordinary event.

Upon returning home, I sent off an email to Alan Combs, who had just written a marvelous book on the subject, *Synchronicity: Through the Eyes of Science, Myth and the Trickster*.[5] Within a few moments, he responded and shared his opinion that my experience with the bird in Mexico was indeed a most remarkable synchronicity. And he offered something further: he told me that when he opened my email, he had been reading *The Sense of Being Stared At*, a book by the revolutionary British biologist Rupert Sheldrake (to which I referred in chapter 2), and happened to be in the middle of a chapter on how animals can predict earthquakes. He suggested that I email Dr. Sheldrake and share this confluence of events with him, which I did. Sheldrake and I then discussed my pursuit of integrating synchronicity into my therapeutic approach, in which I seek to move beyond the limitations of merely rational communication. Collaborating further, we went on to give a talk at Yale University, in which we discussed a convergence of our ideas.

When Carl Jung introduced the term *synchronicity* in the 1920s, he defined it at various times as "meaningful coincidence" and "temporally coincident occurrences of acausal events."[6] You might recall that this acausality was exactly what Einstein was arguing against in his debate with Niels Bohr. Jung was positing that he could find no plausible cause-and-effect factor at work and that such occurrences couldn't be rationalized as mere coincidences. The other word that Jung applied to such an experience was *numinous*, suggesting that a profound experience with deep meaning had just occurred; such an event causes our being to tingle as we move beyond the rational to the transcendent.

Synchronicity poses a major challenge to classical thinking, yet it dovetails naturally with the lessons being learned from quantum physics. Synchronicity occurs when space and time—the basis of causality and separation, for our purposes—retreat, and the universe, our psyche, and the material world appear as one. Jung referred to this as *unus mundus*: "one world." Here we experience the state of entanglement seen in the quantum world manifesting in our macro world, such that separation ceases to exist. The experience of synchronicity may occur on differing levels, yet the theme of acausality remains. Because there appears to be no causal link between events, which seem clearly distinct and disconnected, mere probability can't explain the occurrence. Once again we're experiencing inseparability, but now on an altogether different level.

This phenomenon far exceeds the entanglement of two people and is inexplicable in terms of the mechanistic paradigm. In my earlier example, I could determine no causal link between my writing and the appearance of the bird, yet mere coincidence at the very moment of such intuitive significance would seem extremely unlikely. Two differing states, the psyche and the material world, briefly became as one. Synchronicity removes the division between mind and matter, and when that happens, we start to reintegrate the connection between consciousness and the material world.

Many years earlier, I had been at a yoga retreat center in Massachusetts called Kripalu. As my dear friend Charles and I enjoyed a relaxing evening, he told me that he was considering ending his marriage of thirty years. The next morning he shared the following dream with me: Charles was about to embark on a ferry crossing a sea channel. He had with him a dear and cherished possession—a rare bird. He was fearful that the bird would fly off during the journey, so he placed it inside his eyeglass case and bound the case so the bird couldn't escape.

Charles asked me to assist him in interpreting the dream. I proposed that the bird might have symbolized the flight of freedom, or in this case, his leaving his wife. It was noteworthy, though, that he chose to protect the bird by enclosing it in his eyeglass case so that the bird could not fly off. Was his dream telling him that he'd been choosing not to look at his flight of freedom? We discussed what that symbolism might suggest as we walked off toward the large cafeteria where breakfast was to be served. There was a chalkboard at the entrance to the grand room with a quote or reflection for the day. It read, "Chirp, chirp, chirp. Good morning, little birdie." Was this coincidence or synchronicity, two seemingly distinct realities—consciousness and the material world—temporarily appearing as one?

One evening in New York, one of my Emergent Thinking workshops was exploring synchronicity. During a break, the conversation turned to topics more banal, and we began discussing auto insurance issues for young drivers. As we did, I realized that I had neglected to list my son, a new driver, on my policy. A few minutes later, I noticed that Alex had sent me a text saying that he needed to speak with me, and I shared that message with the group. Three people were sitting to my left and three to my right. Given our immersion in the matter of synchronicity, the group participants assumed that his text related to our discussion, a conclusion that would have appeared irrational in other circumstances.

I can't recall now what my son wanted to talk to me about, but that isn't the point. Oddly, those to my left all thought that my son had just had an accident, and those on the right assumed that he wanted to talk with me about our insurance policy. That night I mused about the differing perceptions of group members. I wondered if it had something to do with where they were seated or if it was simply random chance. The following day, a woman with whom I was working, who had not attended the group the prior night, said, seemingly out of the blue, "Perception is like the dial on the radio. The right side picks up something different than the left side." Such occurrences have been happening with greater frequency as my personal antennae have shifted toward the emergent worldview in which entanglement is no longer seen as a weird thing.

Just as I have been assimilating intellectual intuition into my work as a therapist, I also increasingly seek and experience these synchronistic entanglements.

I had been working with Richard for more than six months, initially focusing on issues around his marriage. As our sessions evolved, Richard began to see that his fear of confrontation stymied his relationships not only with his wife but essentially with everyone. I shared the concept of synchronicity with him, as I felt that something beyond cognitive insight was required to help him overcome his struggles. An avid reader, he mentioned an unusual expression he had come across in one of his readings. The author used the term *inner governor* to refer to the ability of an individual to govern his or her own behavior. This exchange didn't appear to be of any particular significance, and our meeting concluded.

The following week, Richard excitedly told me he had recently had dinner with a new friend, who had told him that while his fiancée was out of town, he had taken some liberties by drinking and partying excessively. "When my governor is out of town," he told Richard, "I'm a free man." Richard was astonished that an unusual usage of the word *governor* had come up for him twice in one week. I asked him what he made of it. He felt that it was no mere coincidence. As we talked further, he concluded that these nonrational synchronistic occurrences were intended to awaken him to how strongly he governed himself so as to avoid confronting others. His inner governor had been ruling him. His strict monitoring of his interactions with others had deprived him of becoming his authentic self.

Since then, Richard has made great strides, becoming more genuine as he loosened his grip and allowed himself to speak his truth. The word *governor*, used far from its typical context on two occasions, helped free him from a habitual behavior that was restricting his life.

Limiting ourselves to rational reflection and consideration about our life is certainly of therapeutic value, but opening to the intuition and how it allows us to interpret the synchronicities provided by the universe is also of inestimable value.

HOW DO WE GET THERE?

When we begin to see in and through wholeness, remarkable experiences present themselves to us. From this vista, we reverse the classical training of our minds that has us seeing only small slices of reality.

The way we think is paramount to this transition. Once you resist binary, either-or thinking and searching for simplified answers to simple questions, you begin to think differently. When our thought ceases to fragment wholeness, the interconnected nature of the universe reveals itself to us, allowing us to fully participate in the reality-making process.

Set your intention to welcome experiences of synchronicity. When these startling experiences occur, it's essential not to default into "That's so weird." Instead, view these experiences through the lens of wholeness. In doing so, you will see that your psyche and the universe have aligned to guide you in a certain direction.

From there, ask yourself what the special meaning is of this kind of occurrence. When synchronistic events occur, profound meaning surfaces as wholeness temporarily supplants the narrower way that we envision reality operating. Wonder, enchantment, and awe replace the causality of the machine.

When our thoughts fragment and dissect, we remove ourselves from all of the wonder and splendor of the world. We construct small compartments in which we place ourselves, and we become imprisoned by this self-deception. In so doing, we suffer from and with the illusion of separation. Learning to cast off the limitations of our constructed reality permits our lives to emerge with purpose. Our sixth sense is there, just beneath the surface, waiting to reemerge. Synchronicity offers a spiritual reengaging with our lost participatory relationship with the universe. We are in correspondence with all, and all is in correspondence with us. When you choose to see in this manner, synchronicities happen with greater and greater frequency. They become the essence of intellectual intuition, guiding us from a deep wisdom.

10

ENTANGLEMENT IS THE
HEARTBEAT OF LOVE

The experience of falling in love is altogether reminiscent of quantum entanglement. Recall that our quantum photons, once having a shared state, are forever connected, no matter how distant they are from each other. The *falling* part of the falling in love process requires a falling away of individual boundaries as the two people merge significant parts of themselves. The coupling moves the two individuals into an entangled sense of oneness.

All living beings are energy fields manifesting through their physical form. Mere physical attraction to another is based on sensory stimulation, but being in lust is not quite the same as being in love. Falling in love requires that our energies coalesce with one another. When this occurs, our energy field resonates with our partner's energy field, and our vibrations harmonize with each other's, so that two individuals are no longer distinctly separate. This energetic interchange happens simultaneously on physical, emotional, and spiritual levels, and it is what makes falling in love—and staying in love—potentially the most fulfilling experience in life.

Over the course of time, however, many people indicate that although they may still love the other, they no longer feel *in* love. There's a common belief that as the years pass, falling out of love is natural and to be expected. I'd suggest that it may be ordinary, but that doesn't make it natural. This retraction from a sense of oneness is often caused by the influence of mechanistic thinking. Falling in love and sustaining it requires maintaining entanglement.

When two individuals are entangled in a state of oneness, empathy is a natural occurrence. Unfortunately, these qualities are unsustainable once we retreat to separation and individualism. As a couple slips back from the endorphin-laden experience of Eros, they start to fend for their individual needs, often at the expense of their entanglement. You should always maintain a healthy sense of yourself within a relationship, but you also need to secure a balance such that your sense of self and sense of relatedness are in equilibrium.

ARE YOU AUTONOMOUS, DEPENDENT, OR INDEPENDENT?

The manner in which I'm applying the words *entangled* and *entanglement* is different from the lexicon's definitions. I'm also not speaking of entanglement as codependence, in which another individual disproportionately informs one's sense of self. Yet falling in love or being in love means that how you feel is greatly affected by the current state of your (entangled) relations. The resonance experienced in such partnering is akin to the state of feeling in love. Eros aside, parents often feel this way about their children. When asked, "How are you?" some parents may at times respond, "I'm as good as my children are." This response speaks of inseparability.

At the other end of the spectrum, some individuals appear to be thoroughly independent and are hardly affected by others. They have sufficiently walled themselves off from emotionally intimate connections as they protect against their insecurities, presenting themselves as impenetrable, and so they are not good candidates to prosper in relationships. They must work through their core wounds—their confining wave collapses—and ensuing defense mechanisms before they can allow themselves to be emotionally intimate.

Their counterpoint would be people who seem overly dependent, so much so that their sense of self is lacking and they are excessively influenced by their need to please—or to avoid displeasing—the other. If you identify with this description, you might be inclined toward what we call codependence. Your goal is to foster the elements of authentic self-esteem, which we touched on in chapter 8, as you move away from seeking other-esteem.

The healthy balance between these two extremes is what I refer to as being autonomous. Such individuals have progressed in their growth and have established a sound relationship with their authentic self, giving them an authentic sense of self-esteem. People in the autonomous category can allow themselves to slip into the entanglement of inseparability without losing an inherent sense of their own self. It's like juggling two balls—one ball representing yourself and the other your partner. You must pay attention to both yourself and the other simultaneously. You are both an individual and inseparably linked to another person. This may sound illogical but participatory relationships require moving beyond the either-or divide toward which fragmented thinking orients us.

ENTANGLED PEOPLE, ENTANGLED ISSUES

Relationships offer a special opportunity for personal growth, though typically not without some disturbances and challenges. Interpersonal relations showcase the chronic issues that each person brings into their partnership, as their individual coping mechanisms and wave collapses spill over into the relationship. Our tendency is to blame each other for relationship problems, which usually results in each person feeling invalidated or treated unfairly. When this happens, we pull back from the sense of oneness and see ourselves as separate and distinct from each other. We begin to differentiate issues as *his* problem or *her* issue. What began as a loving, connected union begins to dissipate into conflict as oneness withers into separation.

I'd been working with a couple in the early stages of their relationship. They weren't married but considered themselves to be in a monogamous, committed union. Judy was recently divorced, with two college-age children, and she maintained a close, if not amicable, relationship with her ex-spouse. As our sessions progressed, it became evident that Judy's need for cordiality with her former husband was masking an underlying issue. She felt compelled to be well thought-of, by both her children and their father, which compromised her authenticity as she avoided any risk of confrontation. Protecting herself from the threat of not being well-liked, she both appeased her ex and avoided providing appropriate parental guidance for her offspring. This provoked her new partner, Howard, who was perplexed by her behavior. He felt as though Judy were more concerned about what her former husband felt about her than by what he himself felt.

Judy's desire to avoid confrontation and to make certain not to displease others had its roots in her own childhood. They were a two-part coping mechanism she created as a child in defense of a wave collapse—not feeling loved by her parents. She believed others would find her lovable only if she accommodated their wishes. Perhaps predictably, she ended up displeasing her new partner at the cost of not upsetting her former one.

Howard came into their relationship with abandonment issues dating back to his mother's departure from his life at an early age. He shared that his fear of rejection had affected his prior relationships as well. Howard tried to assuage his anxiety about Judy's commitment to him by an exhaustive examination of her behavior. Feeling unloved, he relentlessly critiqued her interactions, emails, and text exchanges with her former husband and her children. As a result, Judy felt perpetually blamed. Their relationship was beginning to unravel as they blamed each other for its demise.

When we see our partner's insecurities and challenges as separate and distinct from our own, we are being tricked by the illusion of separation. Their issues become our issues. The issues may be different, but they aren't separate. Picture a drop of ink as it is dripped into a beaker of water. The ink disperses throughout, and you can no longer find its trail. The same thing happens in relationship. Each party's fears, challenges, and unresolved issues become interspersed with their partner's challenges and trigger further reactivity, exacerbating the couple's problems.

As Judy and Howard clamored to have their individual needs met, they became warring individuals, competing with each other rather than tending to each other. This competitive energy was the opposite of the energy they enjoyed earlier in their relationship. Affirmation turned to repudiation, and the ability to listen defaulted to right-versus-wrong attacks. They retreated from their loving engagement with each other into the competitive spirit of individualism. Sadly, these themes are commonplace in most relationships, perhaps more so in romantic partnering but also in platonic relations.

I often hear one person, claim, "I have no issues, but my spouse certainly does." How silly! If you believe your partner has issues, they are sure to affect you—which means that you have an issue as well. Trying to compartmentalize the other and yourself into separate silos is seeing through the lens of Newtonian separation. Two people who are intimately connected must see themselves through the filter of their coparticipation.

Picture yourself on a seesaw with another person. You're up in the air, and the other person is on the ground. Can you move to the ground without changing the other person's position? You are inextricably connected, each person affecting the other. If we move past the transactional attitude that sets up a win-lose, you-versus-me stance into the perspective of entanglement, we can begin to operate from a win-win mind-set. This is at the heart of participatory relationship.

IT MAY BE PERSONAL TO ME, BUT I DON'T NEED TO PERSONALIZE IT

As I continued working with Judy and Howard, I helped them recognize and take ownership of their habitual coping mechanisms, which were exacerbating each other's upset. I worked with Judy to develop a more genuine self-esteem, enabling her to get past her fears about how others saw her. I helped Howard appreciate that Judy wasn't abandoning him but was operating from fear as she avoided confrontation. These problems were indeed personal to both of them, but it was essential not to personalize them.

They each came into their relationship with their personal history. Their confining wave collapses, fears, and insecurities were personal to each of them. And these artifacts certainly had an impact on how they felt about each other. But not personalizing those artifacts meant they did not have any intention to inflict pain on the other. Unresolved challenges can have unintended ripple effects, but it's best not to succumb to the angry emotions that arise when we think, "They are doing this to me." Instead, entanglement allows for the rise of empathy, which can help us remember that we came into a union with another human who is probably still working past the scars of his or her past, just as we are.

I helped Howard depersonalize Judy's fearful attachment to her ex's approval and asked him to feel compassion for her. She was operating from a fear grounded in her childhood, long before he came into her life. It was greatly affecting him, but it wasn't about him. After some effort, this approach succeeded; Howard no longer felt threatened by abandonment and began to lighten up on Judy.

If you find yourself in an adversarial situation with your partner, ask yourself, "Is my partner intending to hurt or devalue me?" You might even ask your partner if it is their intention to be hurtful. If harming you is not your partner's intention, then you shouldn't personalize the behavior.

This is not to suggest that you have to surrender and accept unhealthy behavior. You might say something like, "I feel so unimportant to you when you ignore how I feel or tell me I'm wrong. I feel hurt. Do you care how I feel?"

However, if you're thinking the worst of your partner (because you've personalized their behavior) and go on the attack, you'll trigger your partner's worst reaction, not a connected and concerned response. Whether you choose to connect with empathy or to separate in conflict, you'll get the corresponding result.

SHIFTING THE ENERGY OF A RELATIONSHIP

In the turmoil we experience when a relationship becomes adversarial, we need to acknowledge or change something to shift the energy away from separation and back toward entangled wholeness. Making that shift may mean changing our beliefs, our perceptions, or our behaviors, or possibly all of these.

If you set out to reenter the energy field of the initial romantic entanglement or the caring friendship, you can selflessly try to get in the other's shoes. Doing this doesn't mean you are abandoning your position; it simply means loving and validating your partner. If I try to appreciate and care about my upset partner's point of view, I'm invoking a shift of energy. Connecting empathetically with our partner is the most powerful thing we can do in such troubled moments. It can turn the tide from a competitive—maybe even emotionally and verbally abusive—exchange back into a loving energy field once again entangled with caring. (If you try this approach consistently and with genuine affection, but your partner doesn't reciprocate over time, you might well consider whether the relationship is right for you.)

Another way of shifting the energy of a relationship is to express positive feelings or appreciation for your partner. Once a couple's entangled, participatory energy has drifted into separatism and conflict, they may default to unloading critical thoughts and feelings with each other. Negativity then fills the divide they have structured. Yet there are times in therapy when individuals may share with me positive or appreciative feelings they experienced about their partner. When I ask, "Did you share that with your partner?" it's the exception for me to hear a yes. Why would we become acclimatized to sharing the negative, yet feel awkward or reluctant to express approving or positive feelings? Because we've gotten stuck in the groove of negativity, which only widens that gap between us. We may be holding back an expression of approval so as not to give the other a stronger hand—a sign that we have set up separate battle stations. So set your intention: when you feel good about the other person, articulate it to him or her.

In trying to reset the downward spiral of the relationship cycle, it may be helpful to pause and not be reactive. Take a moment before criticizing or defending and ask yourself, "Does this really matter?" If it doesn't, you can choose to let it go and collapse a very different wave. Again, this is an energy shifter.

From the mechanistic view of separation, the common expression "You can't change the other person" appears sensible when a relationship is in turmoil. But from the quantum view of inseparability, if you change some aspect of yourself, it will necessarily affect your partner, because you're both as connected as our quantum photons. My clients Judy and Howard were both demanding change from the other person, a strategy that rarely succeeds. I explained the concepts of self-change and inseparability to them, both in our couples' sessions and individual meetings. After some time, Howard was able to share with Judy that he appreciated her dilemma and was happy to be supportive of her attempts to break through this old behavior. By making a change in himself, Howard was able to turn the tide of their entangled relationship.

INVITING DISSONANCE INVITES GROWTH

As I've said, relationship provides us with a golden opportunity to grow. Each person's challenges evoke and invite growth in the other. The question is what that growth will look like.

Here we come back to the theme of *being* versus *becoming*. Think of your relationships as catalysts for your process of becoming as you try to move past your challenges. If two individuals see themselves through the prism of becoming, they must see their relations with each other similarly; this is the earmark of a participatory relationship.

One of the fundamental purposes of close relationships is to open our awareness to aspects of ourselves that we need to tend. If we existed in isolation, we'd never come to see aspects of ourselves through the eyes of others. I'm not referring to defaulting to other-esteem, whereby we lose our genuine sense of self, but rather developing the ability to stay grounded in ourselves yet open and receptive to considering how the other person experiences us. Our close connections with others catalyze our further entry into the flow of becoming as we move past our stuck state of being. Ask yourself, "What do I need to see in myself that is being provoked by the other person? And, conversely, what am I precipitating in the other?" Try opening to the dissonance that arises when you look at yourself differently from the way you ordinarily see yourself. If the other person sees you differently from how you see yourself, do you then think the other person is wrong? If so, that conclusion doesn't allow dissonance.

This consideration of dissonance as a positive value opens the door to new possibilities for our personal evolution as we move further into the process of becoming. It also facilitates our greatest potential for thriving relationship. A participatory relationship allows us to see our role in the whole of the relationship rather than as separate from the other person, as if we were part of an equation and each of us were on different sides of the equals sign. The latter is a transactional relationship derived from the thesis of separation.

EMBRACING SUBJECTIVITY AND UNCERTAINTY

One evening, I was a guest speaker for a gathering of people going through the divorce process. After I invited audience participation, one man began railing venomously about his soon-to-be ex-wife. From listening to him, you might well have assumed that she was a hateful, odious, selfish woman. As he continued to vent, we learned that his wife was now living with her new boyfriend. I suggested that her new lover evidently didn't see her in the same negative light. So which woman was she really? Was she a detestable ex-spouse or a loving, devoted girlfriend to her new partner—as her ex-husband had experienced her when they first met and fell in love? It all depends on whose subjective perspective we take and what point in the continuum of the relationship that perspective reflects, as our perspectives certainly change over time. We need to ask whether the other person has actually changed or whether our perception of the other person has shifted. Recalling the problem with binary, either-or thinking, the only sensible answer to this question would be yes. We might certainly agree that the energy has shifted.

Remember that inseparability leads to the loss of objectivity—and that loss is a gain for healthy relationships. Should your partner say to you, "You're not being objective," you might smile and say thank you. The myth of objectivity induces us to separate from the other person and trick ourselves about who they truly are. Yet by acknowledging that reality is subjective, we can come to see how our consciousness and perceptions craft how we view each other. Instead of confusing our thoughts of others with being the objective (literal!) truth, we can view our thoughts as subjective (participatory) representations of them. We then see others—and their actions—as projections of our perceptions of them. Instead of saying, "She is a selfish person," we would do better to say, "I experience her as selfish." This shift of perspective opens the door to seeing what my participation in this process may be. If my behavior angers her and remains unresolved, might she begin to act selfishly with me?

If we see the other person as lovable and nurturing, we will focus on those characteristics. The tendency will then be to filter out or marginalize any perceptions that might be incongruent with the picture that we have created. This phenomenon may well lead to a partner saying sometime later, "He's not who I thought he was." Through this metaphorical looking glass, what we see is inclined to be self-ratifying because we see what we are looking for.

Here is where we deceive ourselves: we lose sight of the fact that our predispositions participate in fostering the reality of the relationship. It also does not occur to us that the other person changes the moment that our perception of them shifts. This occurs in two ways. First, our perceptual shift paints a new figurative picture of the other for us to see. If I look for and focus only on Rosa's wonderful qualities, then Rosa will continue to appear that way to me. Second, if my perception changes and I begin to see Rosa in a critical way, this will, in turn, affect how she sees me. If I'm no longer the same loving and caring person to her, this will clearly induce Rosa to see me differently and alter her behavior toward me. Again, we're both riding the same seesaw, and the energy of our relationship shifts based on our own internal change.

Something similar to Heisenberg's uncertainty principle—by which we cannot know both the location and momentum of a particle at the same time—also holds true for our relationships. What we focus on in the other may well preclude us from knowing a differing aspect of that person. If we feel harmony and acceptance, we're disinclined to judgment (although taken to the extreme, this may lead to avoidance and enabling behavior). Angry thoughts and feelings preclude the possibility of compassion and empathy. The observer effect (discussed in chapter 2) holds that our observation has an influence on what we're gazing at.

Let's also return for a moment to the phenomenon of the wave collapse. You may recall that the act of observation participates in summoning the reality. Prior to the moment of observation, all that exists is potential. Our experience of one another exists in a state of potential until a wave collapse induces us to experience the other in a particular way. Naturally our previous experience and memories of the other individual inform how we will see that person, but our thoughts and feelings will have seismic impact on the energy of our relationship. The ex-wife I mentioned earlier exists in a state of potential. She can be loved or detested. Each observer—the former spouse and the new boyfriend—summons a different reality into the relationship.

Combining the insights of both the wave collapse and Heisenberg's uncertainty principle, we discover that when we see one another in a fixed and deterministic manner, we block the potential of the other person and of the relationship. We can't see certain aspects of the person because they conflict with our preconceptions, and we may artificially "freeze" their identity rather than leaving them the space to change and grow before our eyes—and along with us. In my experience, the most gratifying and longest-lasting relationships are those in which both parties continue to evolve together. They may not change at precisely the same time and in the same ways, but their developments tend to overlap because they both remain open to growth, in themselves and in each other.

Two people stuck in their fixed identity of *being*, relating to each other through traditional rules of engagement and steeped in the illusion of objectivity, face challenging odds for happiness. Yet two individuals devoted to their unfolding process of *becoming*, each responsible for his or her subjective experiences and devoted to enlightened communication, may well bask in the pleasure of the relationship. Learning to engage the potentiality of any moment by not reacting, and so choosing your wave collapse with intention, provides the opportunity for such growth.

THE ART OF RELATIONSHIP

The very manner in which we approach relationship results in many of the difficulties and hardships that we encounter. The conventional professional advice regarding issues of relationship and intimacy—physical, emotional, and verbal—reads like a how-to manual or a prototypical version of a "Six Steps to a Happy Relationship" workshop. Such workshops and seminars abound, offering advice on the dos and don'ts, and pundits enumerate the alleged differences between the genders. Frankly, this routinized approach looks at relationship not as an art form to be cultivated but as a series of steps to master, as though we were assembling a mechanical device.

Engaging each other from a mechanistic stance usually produces abysmal results. Often, people ask me if their relationship is "salvageable." That very question points to the problem of insufficient expectations. We should not seek a repair job or salvage operation, but joy and fulfillment. Fixing and salvaging are the all-too-familiar nomenclature of machinery.

In its ideal form, relationship is a creative, evolving, and beautifully raw experience in which two individuals craft their particular way of communing with each other. Cultivating the relationship is an art that requires sensitivity to the complexity and nuances of two souls engaged in this most important endeavor. This deep, fundamental change in how we view relationship flows, once again, from altering how we see uncertainty. Two people, committed to their process of becoming, can create the opportunity for a joyful partnering when they approach each other from an evolving and participatory perspective. Relationship as a participatory art form is flowing, unfolding, and invigorating as it spirals into deeper and more complex levels of understanding and experience. Just as each individual must engage the process of becoming, the couple must see their relationship similarly. "I'm in a relationship" is far different from "I'm committed to the process of my relationship," mainly because the latter entails becoming as opposed to being.

UNCERTAINTY IS THE ESSENCE OF ROMANCE

As I noted earlier in this book, a primary difficulty in relationships is the inclination toward the predictable routine and the institutionalizing of our interactions, particularly in committed relationships and marriage. The very nature of institutions is that they are designed to specify the rules of engagement and conduct and the expectations of behaviors. They format and regiment us. When we learn the rules of relationship and play by those rules, we become indoctrinated to certainty, routine, and boredom. Over time, this approach erodes the authenticity and engagement that relationships require to thrive. The fresh perspective of the quantum paradigm can liberate our relationships from predictability and stagnation by summoning new potentials.

Oscar Wilde wrote, "The very essence of romance is uncertainty."[1] If this is so, then predictability must be its death knell. As we've discussed at great length, once our habits become routine or predictable, they no longer require us to be fully present because the outcome is known in advance. The formatting and indoctrination of our relationships is counterintuitive and inimical to an emotionally alive and intimate experience.

If we view relationship from the context of the emerging participatory paradigm, we will see it as a vibrant and evolving partnering, inviting both the vicissitudes of change and their ensuing challenges. The vitality of a relationship depends on both parties being open to each individual's unfolding growth along with that of the partnership. I'm not proposing that couples seek an unsafe, volatile experience, but that they endeavor to ride the currents of uncertainty and change, which can propel their individual growth and lead to corresponding growth in the relationship. Embracing some degree of uncertainty is necessary to keep the wind in the sails of the relationship.

So frequently in my couples' sessions, I've noticed that as one person begins to express himself the other begins to react, even if nonverbally. In the midst of a session, Hank began to share some of his perceptions about his wife, Julia. He was communicating in a nonadversarial way, but I nevertheless noticed Julia's face grimace and her body stiffen. After a while I gently interrupted Hank to ask Julia what she was experiencing as I commented on her nonverbal clues. Julia said, "I know what he's going to say before he does. There's no need for him to go on." This level of predictability leaves no room for surprise, wonder, or genuine inquiry. Certainty deadens the ability to be present and precludes playfulness or spontaneity. When I asked Hank to continue, he actually had some unanticipated perceptions to share with Julia, to which she would have otherwise turned a deaf ear.

COMMITTING TO BECOMING

We talk about commitment all the time, but it appears to have different meanings for different people in different situations. The word usually evokes a strong sense of devotion, often accompanied by a declaration about the seriousness of our relationships, such as, "I'm in a committed relationship" or "I'm completely committed to this marriage."

These offerings of relationship commitments are usually statements about behavior or proposed outcomes. They suggest that I won't be seeking another relationship or that I'm going to be monogamous. These pledges are prevalent in many unions, particularly marriages, in which we apply legal vows to substantiate our pledge of fidelity, if not continued love. However, statistics reveal that even when we formalize our relationship through the compact of marriage, we have as much likelihood of failure as success. Infidelity and/or divorce occur at about the same rate as fidelity and lifelong marriage.

Some time ago I attended the wedding of a couple in their early thirties. As I listened to the sermon and the wonderful vows they were making, I mused to myself that in all likelihood they would fall well short of their vows. Not that they would necessarily divorce—there was only about a fifty percent chance of that—but they weren't likely to sustain their loving feelings because they were committing to outcomes. As I noted earlier, they should have committed to the *process*—the process of becoming. This process seeks to further the kind of intimacy on all levels—emotional, verbal, and physical—that leads us to the wondrous, fulfilling relationships we seek. The tenets of the old worldview (separation, objectivity, and certainty) all nullify these goals. They lead to a loss of wonder and ultimate stagnation as we succumb to certainty and further our demise through the individualistic, competitive mandate of separation. As I've witnessed up close in my work counseling couples, when we embark on the path of participatory relationships—imbued with inseparability, uncertainty, and possibility—they can provide an alchemy by which we may achieve and sustain what we yearn for. After all, we just want to be loved, don't we?

11

COHERENT COMMUNICATION

A mastery of our communication skills is essential for experiencing authentic self-esteem, successful careers, self-empowerment, and bountiful relationships. Unfortunately, we are ill-prepared in this realm, given that interpersonal skills aren't held in high enough regard to be part of our educational curriculum. This problem is only made worse by our continued adherence to the mechanistic values of separation, objectivity, and certainty that throw up roadblocks to our success in communicating. In this chapter, I'll demonstrate that by employing the quantum tenets of inseparability, uncertainty, and subjectivity—continuing to integrate the principles of the participatory worldview—we can develop a mastery of participatory communication.

THE FIVE PERCENT RULE

Early in my career as a therapist, I was feeling frustrated in my attempts to assist a couple with whom I was working. They were tirelessly mired in argument, and listening to them was like watching a Ping-Pong ball being knocked back and forth, only no points were being scored. I was searching for a way to help them slow down and listen to each other, to get past their gridlock. In the midst of one session, I reflected for a moment on how I might approach their impasse differently. As I've noted in chapter 9, when I pause, get out of my own way, and set my intention for an insight, it often appears.

I began by asking the husband, Joseph, "Can you try to find just a small percentage of what Helena is saying that you might agree with? Let's look for just five percent you can acknowledge and temporarily suspend the ninety-five percent you're sure she's wrong about." I was asking Joseph to act counterintuitively by neither defending himself nor trying to score a point. I explained to him that he wasn't pleading guilty or surrendering; the goal was simply to establish a rapport so that they could begin to truly hear each other. He finally managed to affirm one of his wife's complaints and took ownership of a particular action of his that continually upset her.

I noticed that Helena barely paused as she was about to go right back into the fray. I raised my hand gently and suggested to her that she reflect for a moment about how it felt to be at least partially validated. Somewhat begrudgingly, she offered to Joseph, "I appreciate your caring about my feelings and acknowledging that you did hurt me." I then asked Helena to validate some part of Joseph's issues with her, and as she did, they began to turn the corner. Their energy coalesced as they moved from the individualistic competition of being right into the collaborative effort to empathize and connect. They began to shift back toward union from separation. A new technique was born for me—one that I now call the Five Percent Rule.

Affirming that there is five percent of a person's argument you agree with in no way means that you have to abandon your position regarding the ninety-five percent you disagree with. You have simply laid the groundwork for the other to take in what you have to say. You have disengaged from trying to win an argument, and a collaborative effort can follow. The Five Percent Rule permits us to halt our addiction to being reactive and move toward being responsive. The success of this new approach reaffirmed for me the superiority of the quantum principles: inseparability seeking rapprochement fares far better than separateness seeking to win; uncertainty (hearing what the other has to say) trumps certainty (knowing in advance what the other will say and repudiating it).

The next time you are engaged in a disagreement or confrontation, challenge yourself to resist the argument and search for a small piece of what the other is saying that you can affirm. Once the other person feels heard and, moreover, validated, he or she may be in a far better position to take in what you have to say.

Timing is essential here. If you rush to reframe or assert your own position, your affirmation appears disingenuous. You cannot just say, "Yes, but . . ." That is part of the process of invalidating. Instead, validate something, pause, and let the conciliatory spirit fill the space that would otherwise be occupied by the noisy back-and-forth of argument. That shift of energy now becomes fertile ground for a meaningful transition and constructive exchange.

Even if you disagree with the vast majority of what you are hearing from the other person, you can ordinarily find some small content that you can acknowledge. We typically marginalize, if not ignore, this small part because our default position is grounded in the right-versus-wrong battle. Recalling our cultural mandate to be right, our thoughts seek to refute rather than confirm. Even though we say we care about each other, we don't act caringly. If you need to "win," that means the other person must "lose." How do you think that works out in relationships?

The success of the Five Percent Rule allows both parties to behave with compassion and empathy, cooperating rather than competing. The goal is not to win but to care. You can immediately apply the Five Percent Rule in your communications with others, whether the person is your intimate partner, a friend or relative, a business partner or colleague, or even an adversary.

What you want the other person to hear is important, but you need to set the stage so they can take it in. From there, a healthy communication might emerge. We must interrupt the compulsion to be right and our default to being reactive.

I've taken pains to explain the value of a simple technique such as the Five Percent Rule because it's a perfect example of how resisting the need to be right can open the door to more effective communications in the quantum spirit of subjectivity and inseparability. Throughout this chapter, I'll describe other similarly simple yet effective keys to the art of successful communication.

SHARE MY MEANING?

To experience truly effective interaction—which is the exception—we need to establish a shared meaning around the words and ideas that we are conveying. We can then further that meaning through a coherent flow of dialogue. Such a skill set allows relationships to thrive, businesses and organizations to prosper, and nations to endure and sustain peace. What could be more valuable?

Coherent communication doesn't require agreement but simply a shared meaning. Coherence suggests that the communication correlates; it matches up in that the other person is "getting" what we're trying to convey and vice versa. We need to know that we are actually talking about the same thing. How often do we pause and thoughtfully ask other people what they mean by the word or words they're using?

One day I was walking by a restaurant near my home and saw a parking attendant named Jacques, with whom I was acquainted. I asked him, "How are you?" He smiled and said, "I can't complain." As I continued my walk, a thought occurred to me; he might have meant either that he had nothing to complain about or that he wasn't giving himself permission to complain. On my return home, I ran into Jacques again and genuinely inquired which meaning he intended. It took some time for him to admit that he believed that no one would care to listen to his complaints, and so he wouldn't bother. That he "couldn't complain" was now clear to me. I explained to him that when

I asked how he was, I truly did care, and perhaps he might make an exception to his rule. Surprisingly, Jacques then shared some significant challenges that he was enduring. If I hadn't bothered to confirm what "I can't complain" meant to him, I would have assumed all was good in his life.

Sharing our meaning is an integral part of emotional and verbal intimacy. To pause and ask people what they meant by the words they've just spoken is also deeply respectful. Respect comes from the Latin *respicere*, which means "to look again." That is precisely what shared meaning demands—looking again at what others intend in their articulation. We need to check in and confirm that we are on the same page that they are.

What someone *thinks* I've said is ultimately more important than what I was intending to communicate. The disconnection between my intended meaning and another's differing interpretation of my words can disrupt the entire exchange. I must be thoughtful and selective in my choice of words so that I increase the likelihood that I'm understood clearly. As Abraham Lincoln said, "We all declare for liberty; but in using the same word we do not all mean the same thing."[1]

We take for granted that our words convey what we intend. In my experience, this assumption is grossly misinformed. As a therapist, I'm still stunned by the impact of miscommunication. Dave and Karen were in a relatively new relationship but finding themselves mired in frequent disagreements. In the midst of a recurring argument, Dave proclaimed to Karen, "I can't do this anymore." Karen became noticeably agitated, and so I asked her what she was feeling. She began to cry and said, "I can't believe he's breaking up with me." I turned and asked Dave, "Is that what you're saying?" Dave looked dumbfounded as he replied, "Not at all. I meant I can't do this arguing anymore." What "I can't do this anymore" meant was clearly open to interpretation.

By the time we exchange a few sentences, a totally misconstrued interaction often prevails. Neither party is sharing the same conversation; their internal monologues have branched off as they react to a word or phrase in a way that the other party may not have intended. This results in a loss of genuine correspondence, which is the heartbeat of communication, and that loss is compounded by the fact that both people may be totally unaware of it. The miscommunication may elicit further damage; feelings become hurt as an emotional landslide occurs. And neither person realizes that they're talking about different things!

In my first session with Jerry and Diane, a longtime married couple, when I inquired as to how I might help them, without pause Diane proclaimed, "He has no idea how to be intimate." Jerry immediately tightened and shot back, "I have no idea how to be intimate? I have no problem at all with intimacy; it's you that does." If I hadn't intervened, they could have reprised a familiar battle and paid scant attention to my presence. They were deeply entrenched in their right-versus-wrong argument but had never identified what they were truly talking about.

I broke in and said, "I'm not at all sure what you each mean by the word *intimacy*. Can you each take a moment and inform each other what this word means for you?"

After a long pause, Jerry explained that intimacy for him ranged from physical affection all the way to sexual intercourse. As he was speaking, Diane looked incredulous. "You must be kidding me! That's not at all what I mean," she proclaimed. I encouraged her to go further. Not surprisingly, she spoke of sharing deep feelings and thoughts with each other in a safe, nonjudgmental way. Once we exposed this essential misunderstanding, born out of miscommunication, the couple was able to engage in a meaningful exchange of their actual needs and preferences.

Words simply represent our thoughts, beliefs, and experiences; they should not be taken as a literal or objective reality. To reiterate: Words don't mean the same thing to all of us. In fact, they ordinarily evoke differing connotations based on each individual's experiences. Our personal history flavors the context of words. We often end up in disagreements without clarifying the things provoking the disagreement. Just consider the confusion around the word *love*. One person says to the other, "I love you." The other responds, "No, you don't." Are they speaking of loving one another or of being in love? Nobody is clarifying what each actually means.

GENUINE DIALOGUE

Sharing meaning is a precursor to a verbally intimate exchange and opens the doorway to genuine dialogue. For example, when Diane accused Jerry of not knowing how to be intimate, Jerry could have chosen to suspend his reactivity for the greater purpose of clarity and compassion. Inquiring what Diane meant by the word *intimate* would have invited understanding. After all, his partner is upset with him. Why not find out what is troubling her? In this instance, he could choose not to be right—by attempting to prove her wrong—but to try to understand what was stirring her emotions. A more considered response might sound like this: "Boy, that feels hurtful. Please tell me what you mean by *intimate* and why you feel I'm failing you." That response might foster a productive discussion instead of breaking down into yet another meaningless argument.

The couple's genuine dialogue could have also been started or furthered if Diane had chosen to model emotional intimacy, thereby inviting Jerry to do the same. That invitation might have sounded like this: "I'm really feeling sad and shut down that you don't share your more private thoughts and feelings with me. I feel like we're strangers just going through life together but not truly connecting. Do you feel the same about me?" Emotional intimacy requires a genuine transparency, in which each person feels safe expressing his or her feelings. When we share our softer side, our vulnerable feelings, we invite the other in. Defending against these feelings in anger shuts the other out and assures our continued upset.

The compulsion to be right, as we've examined throughout this book, is the inevitable outcome of an overly individualistic and competitive—read: mechanistic—culture. As a result of the win-at-all-costs mind-set, we typically misuse the word *dialogue*. A dialogue is far from two or more people conversing about an agreed-upon topic. I'd call that exchange a conversation or a discussion, in which each individual tries to put forth a point of view. This type of interaction plays out on the surface, and little new learning or insight typically occurs. Given a lack of consensus or even a full-blown disagreement, we might expect each person to repudiate any opposing positions. This may be done overtly or in the privacy of each person's thoughts. As we saw earlier, when this happens, the discussion often breaks down into a frustrating back-and-forth in which the words ricochet but no one actually listens. Each party clings to their own subjective truth, but presents it as the objective reality, and the conversation collapses without anyone feeling understood or validated.

By contrast, I would describe genuine dialogue as a shared inquiry, temporarily suspending your assumptions and beliefs—your truth—to further the process of shared meaning. In *Dialogue and the Art of Thinking Together*, William Isaacs makes the point that genuine dialogue is a process of thinking together, rather than thinking on your own and then seeking to convince others that your position is the best and most accurate.[2] A shared inquiry has no opposing sides but instead is an interfacing of two or more people, which demands listening. The Greek root of dialogue is *dialogos* ("speak across"), which implies a flow of meaning. This type of exchange is foreign in our culture because we are so much more driven toward winning—making our point—than sharing meaning and seeking new learning.[3]

The paradox is that the only authentic winning comes from understanding, listening, and validating another's point of view—not from vanquishing it—even if we're not in agreement. If we accept the quantum notion of inseparability, the ensuing sense of entanglement connects us to the other person, and as a result, winning or losing a discussion becomes meaningless, if not mutually destructive. When you step forward in a conciliatory way and inquire as to the other's meaning, you shift the energy of the exchange. The quantum principle of inseparability allows us to live by the golden rule, reframing our communication into a nonadversarial win-win. To quote another renowned American, Thomas Jefferson, "I never saw an instance of one or two disputants convincing the other by argument."[4]

To accommodate this shift in communication, we must upend our compulsion to be right, learn the art of listening, and still the reactive pull of our thoughts.

THERE'S ROOM FOR BOTH OF US:
THE SPIRIT OF GENUINE DIALOGUE

The spirit of genuine dialogue as I've defined it is noncompetitive. No one is trying to be right; instead, we seek to understand and appreciate, which, in turn, ordinarily results in our being understood and validated.

Borrowing metaphorically once again from the uncertainty principle, we can't seek to understand or connect and to be right at the same time; one precludes the other. You must choose which potential you want to summon forth. I have often experienced the instinct to defend myself when feeling falsely accused. It's a knee-jerk response and usually a waste of time. I've learned that I'm better served by trying to appreciate what the other person's experience is and how they came to feel as they do. Then I'm free to try to reframe the issue with the person if I choose to, and we have a better chance for understanding, if not resolution. With the Five Percent Rule, we saw that finding a way to affirm just a small fraction of the other person's argument can help us remain open to communication.

In couples counseling, I find people relentlessly interrupting each other to say, "It wasn't two times; it was three times," or "No, it was six months ago, not four." The corrections are usually just a distraction that interrupts the flow of communication and thwarts dialogue. Taking a departure from the mechanistic either-or, right-or-wrong thinking that plagues us, we should consider the notion that *we're both right*. After all, we both have a right to our feelings and perceptions. A new truth may percolate that allows both people to feel validated, creating room for opposing positions in which no one need be the victor or the loser. This scenario is where breakthroughs occur; they manifest once we engage the complexity that lies beyond either-or thinking, allowing us to learn to think together.

I'd been assisting Andrea and Ted around their loss of sexual intimacy. Ted's previously successful business had recently failed, and he'd assumed many of the household and parenting obligations. Andrea continued to thrive in her career and supported the family financially. They each defaulted to their specific roles, conflict ensued, and their emotional and sexual intimacy withered. Ted felt emasculated and neglected. He'd prepare a well-thought-out dinner for his wife, only to have her pick up a quick bite on her way home, and so he felt unappreciated. As a result, he further lost confidence and stopped complimenting her—let alone making any sexual advances. Andrea in turn felt undesired and ugly and moved more into her masculine energy, directing and controlling many of the couple's decisions.

In sessions, each had been defending their hurt as they lashed out at each other and blocked the other's feelings. I suggested that one person's truth need not preclude the other's truth. What if they validated each other? Couldn't Andrea feel unattractive and rejected and Ted still feel insecure and unappreciated?

As we begin to coalesce in the space of entanglement and inseparability, the distinction between the other and me slips away. A deep awareness of the other is at the heart of authentic discourse and the core of deeper relating. Rather than preparing to defend ourselves, we quiet our voice and let the unfolding process welcome communing on deep levels. This provides an environment in which coherence and cooperation replace the competitiveness and stark individualism of the old worldview.

THE ART OF LISTENING

When I facilitate dialogue groups, we use the Native American concept of the talking stick, the holder of which is the designated speaker. When a person places the stick down, a moment or two must pass for the group to reflect on what the speaker has just said before another person picks it up and speaks. Taking a moment to reflect on what another person has just said, instead of immediately responding to it, is key to the art of listening.

Truly listening to someone is a way of validating what that person is saying. To validate what another has said, as we've seen earlier in this chapter, is to confirm that what you have heard is what they intended to express. Indeed, we don't need to agree in order to validate each other. We simply suspend our opinion temporarily. When we do this, we are genuinely listening, and the other person feels us tuning in to them without judgment.

Authentic listening and validation requires deepening our understanding of the other's feelings beyond his or her words. We must try to tease out the context and personal meaning the person is experiencing and trying to express. This process runs deeper than the simple mirroring technique that many therapists and self-help books advocate, in which you merely repeat back the other person's words, perhaps robotically. We need to move into a shared meaning and inquiry for an authentic appreciation of the other's thoughts and feelings.

We resist validating contradictory positions because we equate this with surrendering our own position—a result of the either-or thinking of the old worldview. But I can care that you feel hurt or upset with me without pleading guilty to the perceived offense. Acting this way is an advanced form of compassion. If I see you walking barefoot and you step on glass, writhing in pain, I wouldn't hold back and say, "It's not my doing." I would care that you're in pain and take action to help you alleviate it. I need to do the same even if you allege that I'm the cause of that suffering.

At first Ted was unsympathetic to Andrea's report of feeling ugly, and he rejected it until I had him inquire further. He asked, "How can you feel unattractive when I desire you so much?" In response, she asked him how he could claim desire for her when he wouldn't even hold her hand or kiss her? They had each cemented their own conclusions, and only through true listening, which invites a shared inquiry, could they come to a deeper understanding.

In the process of listening and validating the other person's perspective, we must share our experiences and subjective truths not as an indictment of the other, but simply as our personal experience. Doing this permits us not only to understand why others feel as they do, but also to advance this understanding into the field of empathy—imagining what it feels like to be them. Getting into another's shoes is both respectful and verbally and emotionally intimate. The person being validated now feels gratified and is prepared to hear what you have to say. From there, genuine dialogue can ensue. This level of engagement is best accomplished by eliminating objective statements of fact and starting your sentences with the pronoun *I*. "I feel . . ." leaves the door open; "you are . . ." shuts it from the get-go.

Speaking in the first person also tends to limit the battle over right or wrong. If someone says to you "I feel hurt" or "I'm feeling angry with you," it makes no sense to defend yourself. If you do, you are denying the other person's feelings and making matters much worse. As noted earlier, the need to defend yourself, be right, or win the conversation precludes validating the other person. Telling the other person that he or she is too sensitive would reflect a complete absence of compassion, let alone empathy, and shows a lack of emotional intimacy. For two or more people to have a dialogue, each has to engage in a new monologue—and learning to listen and validate is one of the first steps in this direction. This process deepens as you search for the part of what someone is expressing that you not only hear (validate) but with which you can also concur. I refer to this technique as *affirming*, which moves beyond the process of validating by affirming that you are not only hearing (validating), but also agreeing with a part of what you've heard.

HOW TO AVOID BECOMING YOUR REACTION

If our goal is to understand and connect rather than win, we must witness and temporarily suspend what our thoughts are telling us. Our thoughts get in the way of our ability to listen. As we saw in chapter 5, old, habitual thoughts, summoned from the archives of every moment we've experienced, incline us toward "re-presenting" the past so that we are not truly present to what another person is saying. To listen closely, we need to note any disturbances created from our thoughts, feelings, and reactions and suspend them for a time. We temporarily avoid taking a position. Recall that we can gain access to the state of potential only as we quiet our thoughts. The superposition—the realm of new possibility—requires liberating yourself from the clamor of old thoughts.

When we react to someone's words without pausing to reflect, we are stuck in the old groove of automatic thought and feeling that we explored in chapter 5. In short, we are not present. We are stuck in the entrainment coming from our old thoughts and feelings. Our thoughts operate from assumption, but we don't ordinarily notice these underlying assumptions. As a result, we're not reacting only to one other person in the moment, but to a complex lifetime of interactions with others and an incalculable number of experiences that lie beneath our conscious awareness.

The distance between the reaction of a millisecond and the thoughtful response of a moment or two can be monumental—like the difference between blasting your horn or tailgating someone who cuts you off in traffic and keeping your hands on the wheel and your foot gently on the brake. That fraction of a moment in which you suspend your reaction—another manifestation of the Possibility Principle—provides the state of potential to facilitate an entirely new and positive wave collapse. This allows for profound breakthroughs in communication and for untold possibilities in relationships.

Learning to observe your reaction helps you not to *become* your reaction, just as seeing your thought allows you not to become your thought. Take anger, for example. If I notice that I'm feeling angry but refrain from actualizing that anger, I can communicate how I feel in an appropriate manner. If I react and become the anger, my angry words and actions are likely to exacerbate the situation and be met with worsening results, which may at times irreparably damage a relationship. My ensuing actions may then shine the spotlight on my behavior instead of the underlying issue at hand. Yet if I say, "I'm feeling angry and want to explain to you why," I increase the likelihood that the other person—especially a significant other—will care about how I feel, and an earnest dialogue is more likely to ensue.

Ted and Andrea turned the corner successfully when Ted got that he could feel unappreciated yet also be open to his wife's feeling of rejection. To do that, Ted had to witness his thought, which was informing him that he was unappreciated. It was just a thought, but a powerful one. He could notice his thought and feeling and not become them.

Dialogue is the language through which this emerging participatory paradigm speaks. Genuine dialogue also insists that we let go of certainty; genuine dialogue requires engaging *uncertainty*, which can enlighten and enliven our relationships. Since we don't know in advance what the other person is going to say, we need not cling to the certainty of our point of view. With nothing new to be learned, a relationship can only stagnate. When two individuals experience a recurring stalemate, with no breakthroughs or shared insights, their communication and relationship begin a decline. Likewise, when members of a corporation or organization regurgitate the same tired ideas and conversations in their meetings, they fail to inspire or be innovative. And when politicians predictably adhere to their same assertions and refutations, stalemates—like the perpetual gridlock in Congress—prevail and we all suffer.

The next time you feel the urge to defend yourself or your position in an interaction with another person, here are some questions you can ask yourself:

○ *What is the other person trying to convey?* Remember that from the new worldview, reality is a subjective experience, so we need to confirm we're on the same page.

○ *Is it more important for me to correct the other person's misstatements or to stay present?* As we've learned, the state of pure potential becomes collapsed with our next thought. Be mindful to select your next thought with care.

○ *Would I rather be right or engage in genuine dialogue?* The mechanistic worldview induces competitiveness while the participatory worldview seeks empathy and compassion.

- *Am I judging or listening?* Watch your thoughts. Are they literal and telling you the "truth"? Recall that objective truths mislead us. Seek participatory thinking that has you take ownership of your thought.

Trying to remain still and taking a moment to ask yourself any of these questions will help you enter the flow of the communication, which allows for genuine dialogue.

THERE'S NO SUCH THING AS A WRONG FEELING

The question about whether feelings can be wrong comes up often in my therapy sessions. The greatest source of invalidation arises from denying emotions, whether we do it to ourselves or to others, or they do it to us. Many people struggle with the question of whether their feelings are right or wrong. Feelings are neither right nor wrong; they simply exist.

I've heard people say, "It's illogical to feel that way," which is an absurd statement, unless you're Mr. Spock on *Star Trek*. What you feel is what you feel. Feelings should not be subordinate to logic and should not be subject to its sovereignty. That said, it's helpful to consider why we feel how we do without judging the emotion. We are free to reframe our feelings by providing context, but we should evaluate our emotions rather than judge them. A softer gaze, rather than an incisive analysis, typically serves us better.

For example, you might ask yourself where your feeling is coming from. The feeling is informing you of something, perhaps historical or unresolved or, alternatively, trying to clue you in to your genuine state in the moment. Take a few seconds and try to differentiate whether the feeling is reflexive and habitual, or appropriate in the present moment. If it's an old feeling coming up far from the context of the current experience, you might choose to release the feeling. You might also employ the exercise of searching for the thought that most likely triggered your feeling. If it's about another person, for instance, it might sound like this: "I can't believe he's so insensitive." You then feel angry in correspondence with your literal thought. Shift the thought into the participatory, "I'm having a thought that he is so insensitive" or "I see him as uncaring," and you reduce the reactivity of your feeling and can then communicate it far better.

Feelings are ordinarily informed by our thoughts, and it helps to explore what thoughts may have triggered what we are feeling because thoughts and feelings tend to act in concert with one another. Again, don't judge your feelings; just note them and then self-reflect. Although feelings can't be wrong—you're feeling what you are indeed feeling—they can be reconsidered. Feeling angry and reflecting on why that is so provides you with space to understand the feeling rather than react from it. You are then more prepared to articulate your thoughts and feelings attentively.

Some people invalidate their own feelings by not trusting them. Such people are particularly prone to having others further this invalidation by their critical comments. Accusing someone of being overly sensitive undermines that person; it's a barrier to emotional intimacy and really a statement about not caring how the other feels. When you're feeling hurt by what someone has said or done, it's incumbent on you to share how you feel and inquire, "I'm really feeling hurt at what you just said to me. Was that your intention?"

SILENCE IS LEADEN—NOT GOLDEN

Over the many years I've been practicing therapy, I've found that couples who are struggling in their relationships often succumb to silence. Sometimes it's one person who defaults to not speaking, and other times it's both. In either circumstance, such prolonged silence—not a healthy pause or meditative break—speaks to the absence of verbal and emotional intimacy. Unless we're communicating on levels of extrasensory perception or body language, words are the fundamental tools available to us to communicate and resolve our issues. Being in a relationship and resorting to silence makes little sense. Not only does it contribute to a decline of the relationship's positive energy, thus sabotaging the lifeline of a healthy coupling, but it also chokes your expressive needs. Silence can lead to despair and depression as our voice becomes stifled.

Telling the other person what and why you're feeling what you are may lessen your reactivity. When you can express what you're feeling—in the moment that you're experiencing it—you're much less likely to act out that negative feeling. Problematic feelings that go unexpressed tend to percolate and boil over; they take on energy of their own, and the ensuing conflict hours or days later may have little correlation to the original emotional grievance. When this occurs, you have less chance of being validated because there may be little correspondence between your hurt feelings and your actions in the present moment.

When we don't share our thoughts and feelings with each other, we are often really trying to control the other's reactions and behavior. If others don't know what we're contemplating, then they can't possibly respond. At other times, we use silence to punish. Silence can be more hurtful to people than angry conflict because it precludes actual personal engagement.

At times, people who are inclined to please others or avoid confrontation fall prey to the negative power of silence. The tendency is to choose silence rather than upset the other party. This propensity often suggests a degree of codependence because one person can't tolerate the other being upset and so guards against that by choosing silence.

When we resort to silence, we create an internal monologue, typically ascribing to others our projection of how we assume they would respond if we shared our thoughts with them. We may play out an entire script in which another person's role is predetermined. We're controlling against unwanted consequences. In doing so, we are locked into a state of stagnation, the communication stalls, and the relationship—whether personal or professional—has little chance to evolve. Then, with no opportunity for resolution, let alone growth, the relationship begins to wither.

I had been working with Anne, who went to great lengths to avoid displeasing her husband, Monroe. She desperately wanted his greater engagement with her, and she found him to be easily upset and derailed. So she deliberated at length about what she'd feel safe sharing with him. Monroe, when upset with Anne, could go for days in silence as he punished Anne for her transgression. Silence of this type is intended as punitive and, as noted, percolates to a heightened intensity for both parties, as they have little opportunity for resolution. In both circumstances, silence is thoroughly nonparticipatory and retreats fully into the separation of the mechanistic paradigm.

Just as expressing one's voice is life-affirming, manipulative silence is soul-defeating. People who default to silence may claim, "He won't really listen" or "She will only throw it back at me, and I don't want to fight." Although this thinking may be understandable, it is self-injurious. We invalidate ourselves when we shut down our own articulation. Fortunately, learning how to be heard is a skill that can be acquired, and there is an approach that can improve our chances of being heard.

When I prepare to articulate something that I believe will be difficult or challenging for the other person to hear, I find that devoting a few sentences as a preface to the intended exchange can increase the chance that I'll be heard. I set the stage so that my words won't fall on deaf ears. Runners stretch before they run, pitchers warm up before facing their first hitter, and we study before taking a test. Just as importantly, we need to ease thoughtfully into a challenging exchange. It may be as simple as saying, "I have a problem, and I'm wondering if you can help." Or "I'm confused about something. Can you try to help me understand?"

"Framing" your remarks with a brief preface such as this, rather than abruptly jumping into a provocative topic, opens the door to a shared inquiry and reroutes you from the anticipated roadblock. When we do the latter, the other party may not be able to truly listen. The other's tuning us out may be triggered by specific words or topics or our tone or body language, but it is probably anchored in the memory of previous impasses and unresolved conflicts. More often than not, others may appear to be defending their territory and preparing their rebuttal while we're still trying to articulate our thoughts, and vice versa. Your first sentence may not be complete before the other person's reaction has begun.

By instead saying, "I'm struggling with something, and I need your advice," you set up the possibility for constructive dialogue. The other person might then say, "Sure, what's the problem?"

An appropriate follow-up might look like this: "Well, I have something to share with you that I really need you to hear. But I'm anticipating that you will just shut me down and tell me that I'm wrong, so I don't know what to do."

You now have a far better chance that the other individual might be less reactive and more present, not to mention sympathetic to listening to your feelings. What we're doing here is acknowledging our history of failed communication and demonstrating our sensitivity to that prior to engaging the actual subject matter. Having properly set the table, we can move into the actual content.

Avoid beginning your preface with the word *you*. If you begin in the first person, the other person may still be listening. Steer clear of the other's defensive reaction by saying how you feel, as opposed to giving an objective indictment of what the other person did or didn't do.

To have a true dialogue or a purposeful exchange necessitates creating *two* new monologues. I also suggest taking a moment before you even begin to speak to ask yourself how you can best articulate what you wish without turning the conversation adversarial. In the following chapter, I'll introduce an entirely new way to use language in our communication that will tie together many of the key elements of this book, providing you with the final elements in your shift into the participatory worldview of possibility.

12

NOT JUST SEMANTICS

Why has our transition to the participatory worldview taken so long? After all, nearly a century has passed since the initial discoveries in the field of quantum physics altered our conception of reality. What holds us back?

I propose that our semantics hold us back. The manner in which we think and speak still relies on words that mire us in classical Newtonian and Aristotelian thinking. This problem appears deeply rooted; our words anchor our thoughts, perceptions, feelings, and relationships in a fixed, deterministic, and objectivity-based language. As the linguists D. David Bourland and Paul Dennithorne Johnston proposed in *To Be or Not*, "Language lags behind the current scientific understanding of reality."[1] As a result, our minds don't sufficiently evolve alongside science's new discoveries; they remain imprisoned by the words that constrain us. George Lakoff, the noted linguist and cognitive scientist, explains it this way: "Language that fits that worldview activates that worldview, strengthening it, while turning off the other worldview and weakening it."[2]

Our words constitute the vessel through which our thoughts and feelings take shape. What begins as an instinctive impression or the visceral bubbling up of a notion takes life through words, and these words in turn shape how we see and engage life. The way that we believe the universe works, how we feel about others and ourselves, our beliefs and our relationships, all take form through our words. And as it turns out, our language imprints our brains. The pioneering work of linguist Benjamin Lee Whorf suggests that by altering our use of language, we also revise how we perceive reality operating.[3] Our transition into the participatory worldview falters when we continue to employ words that reflect the dogma of an entirely antiquated worldview.

"TO BE" PRECLUDES POSSIBILITY

Alfred Korzbyski first proposed eliminating the *to be* verbs (*am, are, is, was, were, be, been, being*) in his groundbreaking book *Science and Sanity: An Introduction to Non-Aristotelian Systems and General Semantics*.[4] David Bourland furthered this position in a prescriptive movement to remove the *to be* verbs, which became known as E-Prime language. Why would we want to eliminate such seemingly fundamental verbs from our vocabulary? Because when we use the *to be* verbs, we objectify our perceptions and make them absolute and fixed, per the mechanistic world. These verbs constitute the building blocks of literal thought—the thought that tricks us into separation, objectivity, and determinism—and stand in stark contrast to language that evokes the participatory worldview.

For example, look at the statement "Joe *is* lazy." This represents a fixed, objective statement of fact. Perhaps Joe *appears* lazy when he feels uninspired or bored or distracted. Or maybe Joe overeats, which makes him fatigued, so he seems lazy. If we extract the word *is*, we might say, "Joe seems lazy today," which provides context, or "I always see Joe as lazy," which accounts for my subjective thought about Joe. "Joe *is* lazy" doesn't sound relative or conditional. Could Joe appear lazy now but highly motivated at a different point in life? If Joe finds his passion next year, will he still appear lazy? The word *is* and all the *to be* words preclude change and suggest that a state of permanence can actually exist. Yet these words make up the linguistic foundation of the mechanistic paradigm. No wonder we struggle with change.

Let's look further at how these verbs preclude the change process. Tom's friend might say, "My friend Tom *is* an addict." If we remove the word *is* and say, "Tom struggles with addiction," we don't fall prey to making an absolute and unchanging statement. In the parlance of Alcoholics Anonymous, one says, "I *am* an alcoholic." This attitude may arguably help people maintain sobriety, but it severely limits their personal and spiritual growth since they must always see themselves in the same way. Ten years from now Tom might proclaim, "I suffered with alcoholism for many years, but I no longer do. Nevertheless, I choose to remain sober." That statement allows Tom to evolve, enabling him to come out of victimhood. Because *to be* verbs deny change, they block us from defining moments. Instead, they cement our realities and obscure our ability to see our participation in the flowing nature of reality.

Remember, the universe appears in a perpetual state of superposition—pure potentiality—in which everything flows. So conditions that appear fixed should seem incongruous to us. We shouldn't say, "Joe *is*" or "I *am*" or "you *are*," for they suggest immutability. The word *is* precludes possibility; *to be* verbs appear antithetical to the Possibility Principle.

The *to be* verbs block us from seeing life as a flowing and unfolding experience and so act in an obstructionist manner. They blind us to movement as they root us to concrete and fixed notions and a determination to keep things static rather than flowing. *Is, am, be* all evoke fixed, inert states.

Heraclitus, the famed Greek philosopher who said, "You could not step twice into the same river," believed in inexorable change.[5] Years ago I often found myself saying, "The only constant in the universe *is* change." I now appreciate two errors in that phrase. My use of the word *is* appears inconsistent with my current beliefs. Further, I came to appreciate that the word *change* has no meaning without the absence of change. Yet as we've seen, everything flows. Our use of the word *is* stops flow in its tracks. Nothing simply *is*.

RELEASING CONFINING WAVE COLLAPSES

Using the *to be* verbs in a negative, self-referential way often marginalizes our self-esteem. "I *am* a loser" seems far more damaging than "I feel like a loser." After all, if I feel like a loser, I might ask myself why I feel that way, how I came to that belief, and how I might change it. My feelings and perceptions can change, of course, but if "I *am* a loser," that quality appears fixed, as a constant.

Let's consider the common refrain "I *am* not smart enough." It would seem far healthier to say, "I feel stupid when I don't know the answer," so that we can permit ourselves to move out of this negative refrain. The *to be* verbs tend to lock in our confining wave collapses, keeping us stuck in the habitual refrain. Indeed, many of these confining wave collapses came to us by using *to be* verbs. Consider the difference between saying to your child, "I *am* so disappointed in you," or "I feel so disappointed for you."

While teaching my Mastery of Thinking course one evening, we explored the wave collapses that informed our identity. I asked everyone to reflect on their childhood, to think about a confining wave collapse and their ensuing belief about themselves. A man in the group who had never previously spoken decided to contribute. He said, "I *am* nothing; I *am* empty." Everyone appeared rather stunned by his vulnerable candor, and a few moments of silence ensued. I then asked him to voice the same statement without using the *to be* verb—without the word *am*. He responded, "I feel like nothing; I feel empty." As he spoke, his face lightened because he realized that his situation could change. What he felt—although deeply embedded over the course of a lifetime—now looked potentially alterable.

You may recall that in my narrative about Sam (chapter 7), who suffered from feeling depressed his entire life, he related how the diagnoses meted out to him proclaimed that he *was* a depressed person. He experienced his life according to his diagnosis. His transition occurred by switching to the thought "I feel depressed" or "I always feel depressed." He could then examine why he felt this way and open the door to shifting his perception. Eliminating the *to be* verb enabled him to see how his childhood wave collapse informed his sense of self.

In another setting, I worked with a gentleman who had spoken of his pessimism. "I *am* a glass-half-empty kind of person," he claimed. His negative and measuring self-reference had imprinted his self-image and his life. I asked him to reframe his thought without using a *to be* verb. When he said, "My thoughts have me see the glass as half empty," he opened to the notion that his thoughts instructed his reality, which he kept locked in by using the *to be* verb.

Reflect on any belief you cling to that limits your life. Try restating it without using any form of the *to be* verbs. Notice how doing so frees you from your belief.

WHOLENESS VERSUS FRAGMENTATION

The use of the word *is* resonates with Aristotelian philosophy, which, as you may recall, proposed that something either *is* or *is not* true. This led us to either-or thinking. Recall that in quantum physics, light has a dual capacity. Because it has the potential to manifest as a particle and/or as a wave, we cannot say light *is*. And as we've learned, our macro everyday world looks thoroughly quantum. The word *is* divides and fractures reality into *is* or *is not* compartments. This simple two-letter word significantly influences the way we think as it fragments and ruptures wholeness into small slices. In many cases, the word *is* imposes a misinformed reality and greatly hinders our communication with both others and our self.

In chapter 6 we discussed the nature of either-or questions that typically require a *to be* verb. As George W. Bush asserted during his address to a joint session of Congress and the American people on September 20, 2001, "Either you *are* with us, or you *are* with the terrorists" (italics mine). Notice the sharply drawn distinction.

To be verbs also reduce complex matters into an oversimplified and wrong-minded choice—something *is* correct, or it *is* wrong. You *are* this or I *am* that. Context and relativity become lost when making *to be* statements. By using *to be* verbs, we paint over the complex, multilayered tapestry of the human experience with broad brushstrokes.

DO FACTS EXIST?

When I discussed this language issue over dinner with friends one night, their ten-year-old daughter, Florie, asked, "If we don't use the word *is*, doesn't that mean that facts don't exist?" This question struck me as wonderfully insightful. She spoke directly to the point. What we call facts simply portray our representations of everything with which we consensually agree. Over time, the facts seem to keep changing. Yesterday's truth becomes tomorrow's folly. The definition of gravity changed from Newton's to Einstein's when scientific observation affirmed his theory of relativity. What happened to the fact of gravity?

In my childhood, teachers and textbooks instructed us that there *are* nine planets in our solar system. Did this represent an unalterable, unchanging fact of the universe? The word *are* would suggest so. New thinking and discoveries inform us that Pluto no longer qualifies as a planet. So facts do change! Indeed, the very word *fact* makes little sense.

With Einstein's theory of relativity, we've begun to see that even time *is* relative. We ought to say that time *appears* relative because the occasion may well come when new insights will once again alter how we see time. But our words trick us into forgetting that. The *to be* verbs trick us through objectivism (turning reality into things) and obstructionism (impeding the inexorable flow of the universe). As a result, these classically oriented verbs imprison us in the doctrinal beliefs of the old worldview. They speak to Newton's world of fixed things.

Our dictionaries—rooted in the concept of definition—further the notion of *is* by the very act of defining. Imagine a dictionary of the future that embraces the participatory paradigm. It might begin this way: "This dictionary describes our consensual understanding of the words that follow. We acknowledge that, over time, these descriptions may change." Such a preface moves from the construct of immutable facts toward a temporal description, which ideally conforms to the participatory worldview.

RELATIONSHIP RESCUE

When we remove the *to be* verbs from our communication, we cease making objective statements that trigger so much conflict in our relationships. We take ownership of our feelings as well. For example, "Bill *is* a selfish husband" sounds vastly different from "Bill seems so self-centered to me." Recalling the narrative about the individual who proclaimed that his ex-wife *was* heinous, note that she *wasn't* always that way to his mind or to her current boyfriend. *To be* verbs present themselves in opposition to the spirit and the practice of dialogue. They thwart us from having generative and shared inquiry, and they move us into opposing camps from which we each cling to our facts. In short, these verbs incline us toward an adversarial energy and lead us to argue about the truth, as opposed to speaking my truth, which doesn't preclude your truth.

Over the last few chapters you've learned the benefits that come from moving beyond the myth of separation toward the possibilities inherent in entanglement. The *to be* verbs sever connection, let alone inseparability. They construct the divide between you and the other. "You *are* so reactive and angry" proposes that the speaker of that thought has nothing at all to do with that observation. Sentences with these verbs speak a nonparticipatory mind-set. The compassion and empathy that we seek from one another remain elusive through our use of these verbs. Eliminating them ushers us into the arena of connectivity, compassion, and empathy.

Removing *to be* verbs from our personal lexicon helps us create authorship of our statements. Rather than saying "You *are* . . . " or "It *is* . . . ," we can say "I think . . . " or "I see it as . . ." This allows us to observe our participation in constructing the truth—or the reality—of what we experience and see. When we reflect that reality in our words, we gain clarity, and we create a more reflective world. In effect, we take responsibility for our words as representations of our participatory thoughts and beliefs as opposed to making authoritative (read: *literal*) statements about the truth. Our personal conflicts tend to engage around the battleground of what *is* or *is not* true. E-Prime language invites us into the heart center of relationships as we express feelings, perceptions, and vulnerability—the manna of connected relationship and effective communication.

Eliminating *to be* allows a wonderful transition to participatory and subjective thinking. When we begin a statement with "I find . . . ," "I experience . . . ," or "I believe . . . ," several things occur. First, as I've noted, we avoid the pitfall of right versus wrong. I learned years ago not to say, "It *is* hot in here," because I often heard back that it *was not* hot in here. When I learned to say, "I feel hot," a dispute couldn't arise. When I express what I feel, think, or experience, my words no longer appear subject to the right-or-wrong test. Second, beginning statements this way leaves the door wide open for others to share their subjective experiences or impressions. Subjectivity presents a gateway to emotional and verbal intimacy, whereas objectivity—represented by *to be* verbs—precludes relatedness. The *to be* verbs speak the objective language of the things of Newton's machine, and eschewing them allows us to sidestep the battle over right or wrong.

BREAKING FREE FROM VICTIMHOOD

As we've seen, viewing reality as "out there"—from the classical worldview—leaves us as dispassionate and removed observers in our lives. From this orientation, we tend to see ourselves as victims of circumstances beyond our control. "I *am* a loser" can transform into, "I feel like a loser when I don't meet my goals." Speaking in this manner enables us to participate in creating our reality. The former represents a statement of fact; the latter speaks of my feelings, perceptions, and experiences, which evoke the participatory paradigm.

The advantages to this linguistic transition appear far greater than mere words can express. You may recall that earlier in the book I proposed that asking the question "Who *am* I?" seeks a fixed answer, which constrains our sense of identity. Remove the word *am*, and you might ask, "How do I see myself?" or "How would I like to see myself?" or "How would I like to experience my life?" Notice the shift from object to movement, from fixed to flowing, from being to becoming. As I delineated in chapter 7, the core of our biomedical, pharmacological, diagnostic attitude appears deeply rooted in the *to be* verb: "John *is* bipolar." Once considered, this statement sounds truly absurd. It turns John into a thing. John no longer appears as a father, a son, a husband, and a friend. The word *is* reduces John into one oversimplified aspect.

Using that form of the *to be* verb speaks in gross generality as it obscures all contextual reference. Worse still, the thing—bipolarity—simply represents behaviors that we believe comply with words that we created! Recall Alfred North Whitehead's theory of misplaced concreteness. We turn the idea—bipolar—into a thing, and then we turn John into that thing as well. This sleight of hand—reification at work—requires the *to be* verbs because once we vanquish them, our sentences force us to acknowledge our mind's complicity in what it created. For example, consider, "When I spoke with John, his behavior seemed consistent with what we call bipolar." That sentence appears alive with subjective context and ownership of my perceptions and, so, opens the door for a new mastery of my thinking. E-Prime can assist in freeing us from the torrent of old thoughts.

FINDING YOUR VOICE:
EMBRACING UNCERTAINTY ONCE AGAIN

Many people feel tentative, if not outright insecure, about sharing their thoughts and feelings. Sensitivity to others' opinions—a symptom of a fragile sense of self and indoctrination in other-esteem—leads such individuals to hesitate. They may vacillate, debating whether to speak, and then question what to share if indeed they choose to make themselves heard. This insecurity interrupts our natural process of articulation, and many people tease out entire sentences in the safety of their own minds before they begin to speak, breaking any sense of spontaneous flow. I think of this halting approach as the equivalent of a silent stutter. Such individuals remain rooted to the principle of certainty.

The *to be* verbs make statements of certainty, with no room for uncertainty. They speak the ideals of determinism. We've seen the extensive damage that occurs from our attachment to certainty and the numerous benefits derived from inviting uncertainty. We must learn to open ourselves to thinking and speaking in E-Prime to embrace uncertainty.

One individual with whom I worked shared a wave collapse cemented by her father's words: "Think before you speak." Those words constructed a fearful image for my client, inclining her to worry about making a "mistake." I have found, however, that using E-Prime language can dissipate this fear.

At first, it may feel awkward to speak without employing the *to be* verb. We need to pause and find a new voice as we struggle to break free from speaking in the fixed, definitive way that this verb mandates. And yet, this reflection helps us take ownership of our thoughts and feelings because we now express our subjective perceptions. When we do, we discover that the fear of appearing wrong or of another correcting us vanishes. When you speak from the first person in a subjective voice, you can feel safe from others correcting you. You can have perceptions or understandings that differ from those of others, but your vantage point should no longer feel subordinate to theirs.

For this reason, E-Prime also serves to build our self-esteem as our feelings, thoughts, and perceptions seem further removed from judgment. Many people who have had difficulty speaking with conviction, whether because of self-esteem issues or fear of not having the facts correct, feel free to assert their thoughts and feelings with confidence when speaking E-Prime. I have heard many individuals whom I have coached in this process acknowledge the benefits as they begin to find their voice. They become comfortable beginning a sentence without making any plan as to how it will conclude. The experiences of these people stand in stark contrast to that of the girl whose father told her, "Think before you speak."

When you give yourself permission to speak in this manner, one word will inform the next, allowing your voice to emerge as your communication takes on coherence. You have to trust in the process. This manner of speaking embraces uncertainty and inclines toward a more generative means of communicating, which stands in stark contrast to the objective, punctuated style that occurs when using forms of the *to be* verb.

The *to be* verb traps us in a rut of old thought, replicating the groove of old confining wave collapses and beliefs. Removing *to be* from our mental map enables the transition from being to becoming, from fixed identity to a flowing, participatory experience. Releasing the *to be* verbs ushers in our transformative process.

A language rooted in the old worldview acts as a roadblock, restricting our more expansive participatory engagement. When we communicate free from these verbs, our thoughts no longer trick us into a false objective reality; we take ownership of our thinking with the advent of participatory thought. When we begin to modify our words and language in such a manner, the way that we experience our lives shifts from automated to present, from victim to author. These semantic shifts appear pivotal in creating a present that becomes unburdened from the past. They also free us to join in with the flow of the universe. Our words shape our thoughts and our world. By releasing those words that constrain us, we foster the space for the emergent quality of the new thinking that we seek.

Please note that, with the exception of italicized words, I wrote this chapter without using *to be* verbs. I did so to provide you with the experience of thinking and speaking without the *to be* verbs. I don't intend this as an absolute prescription for your communication. I wrote all the preceding chapters using these verbs, and there are some exceptions to this rule, namely, using the *to be* verbs as auxiliary or helping verbs, as in "I am driving." I share the E-prime language as an additional tool to use in your movement toward bringing ever-greater possibility into your life.

EPILOGUE

GETTING FROM HERE TO THERE

When we equate the human experience—present and past—with the term *human nature*, we make a misguided assumption. This belief is akin to looking at the behavior of a child and assuming that his or her growth and development had culminated and would experience no further maturation. We may still be in the late childhood or early adolescence of what it means to be human. So our human habits this far become mistakenly commensurate with our human nature. I contend that we are just now on the cusp of venturing forth to embrace our genuine nature, one that thoroughly resonates and aligns with the participatory universe in oneness. Realizing this nature allows us to touch our true potential, heralding our personal transformative process.

Although the notion of gradualism—slow and steady progress—is generally accepted as a cultural meme, it misguides us. Why slow and steady? That approach may be beneficial in running a marathon, but I see no virtue in delaying our emergence into our fuller participation with life. I am opposed to gradualism in general and advocate for catalyzing your defining moments. Why delay your prospering? Why not break through the harness that restricts you? As you've seen, in the instant of thinking and perceiving differently, we open the floodgate of new possibility into our lives.

I've witnessed many people reach these breakthrough moments and sustain them. That is the core of a defining moment. If you don't sustain the breakthrough, it's not a breakthrough. The way to sustain your defining moment is to keep your foot on the accelerator. Yet the question remains, why do some of us venture forth in our change process while others who seek it remain frustrated? And how can you, as an individual, hasten your personal unfolding?

WILLFUL INTENTION

To access and manifest new possibility first requires your intention to do so. And yet, in my experience, many people indicate their intention but don't make the necessary strides forward. That's because intention alone is insufficient. To move forward, intention needs to couple with will. Think of a sailboat. You hoist the sail, which is the equivalent of your intention to move. But without any wind, you're not likely to go far as you simply drift. Think of the wind billowing through the sail as commensurate with your willfulness. Typically, the word *willfulness* evokes a pejorative quality that speaks to one's stubbornness or unwillingness to change directions. Yet this is exactly the quality of resolve and determination that I'm advocating.

What separates those who provoke change and summon new potential from those who continue to struggle is single-minded willfulness. You need to infuse your intention with your unwavering conviction. Intention + Willfulness = Your Potential. When you integrate your new learning and skills and fortify them with your willful intention, you'll be fully participating in your life's experience as you navigate your own course. If you allow hesitation to distract you, you'll be likely to falter.

Our core beliefs and ensuing thought patterns instruct us as to our willful intention or lack of it. They may speak to us as either, "Why I can't" or "Why can't I?" The former obviously expresses your thought looking for reasons not to succeed. The latter clearly opens the door to your future. Pay close attention to which camp your thoughts fall into, and if it's the former, ask yourself, "Where does this belief come from?" You now know the process you must engage to release that belief, that imprint of a confining wave collapse. Seeing the source of what's holding you back allows you to move forward toward the latter question: "Why can't I?"

RELEASE DOGMA AND
THROW AWAY THE RULE BOOK

From the participatory perspective, in which consciousness and the universe are as one, the notion of immutable laws becomes untenable. The construct of an absolute, unchanging reality appears inconsistent with a perpetually unfolding reality-making process. Nothing remains unchanged. As reality perpetually unfolds and emerges, it is not subject to deterministic laws. So how can we have laws of the universe?

The construct known as law is made by humans; recall that reification is a manifestation of our consciousness organizing the world and reality for the purpose of our comprehension. Laws are creations of mind, nothing more and nothing less.

For us to project laws onto the cosmos seems like an excellent example of anthropomorphic tendency. Similar to what we examined with facts and definitions, asserting the "truth" of fixed laws denies the evolving, participatory nature of the universe and depicts our need to categorize and classify. This may increase our knowledge base, but it tricks us into false realities.

We do something comparable through the memes we construct about how we should live, relate, and communicate. This virtual rule book has us protect what feels vulnerable and what causes us to isolate and "win" rather than relate and connect. The rules of life by which we operate clearly fail us. If these operating rules aren't succeeding, why should we remain wed to them? During halftime in a football game, if one team is down by a score of 24–0, you can rest assured the coach whose team is trailing will create a new game plan. That is precisely what I'm encouraging you to do. The existing game plan induces much of the despair we encounter. So let's adopt a new strategy for life, based on the wondrous messages coming from the new science that nurture our human qualities. This domain promotes our authenticity as humans, no longer imprisoned by mechanism. Your new philosophy for life must be devoted to your emergence into full participation in the unfolding—*becoming*—process of your life for yourself and those you touch.

ACKNOWLEDGMENTS

'd first like to thank my personal editor, Peter Occhiogrosso, who has been my most trusted ally in this endeavor. Many thanks as well to Stan Friedman, my "consigliere," and to my literary agent, Felicia Eth. I express much appreciation to the Sounds True staff and in particular their editor, Amy Rost.

My work—this book and my approaches—have been developed through a transdisciplinary attitude in which I looked toward fields outside of psychology to garner a wider gaze toward my emerging ideas. Those who have so inspired and enlightened me include the physicist David Bohm and his monumental work on thought and dialogue. The philosopher Alfred North Whitehead's work prompted radical new perceptions for me. The insights of Ervin Laszlo, Menas Kafatos, and Larry Dossey in the realm of entanglement and oneness have illuminated my path, and so, many thanks to them.

My deep gratitude goes to my partner, Leslie, for her patience, devotion, and encouragement through my writing process, and to my son Alex for his tireless and loving support. To my son Jesse, whose acumen and diligence were so essential in the early stages of writing this book, I say, "Couldn't have done it without you, son." A special thanks to my cousin Eugene, who catalyzed this process by insisting that I write this book. Last, my immense appreciation to my departed parents, Ruth and Sidney, who instilled in me the conviction that all things are possible.

NOTES

Introduction

1. Fritjof Capra, *The Turning Point: Science, Society, and the Rising Culture* (New York: Simon & Schuster, 1982).

CHAPTER 1. The Promise of the Quantum Worldview

1. René Descartes, *Discourse on Method and Meditations on First Philosophy*, 4th ed., trans. Donald A. Cress (Indianapolis: Hackett Publishing Company, 1999).

    For instance, Descartes writes:

    And as a clock composed of wheels and counter-weights no less exactly observes the laws of nature when it is badly made, and does not show the time properly, than when it entirely satisfies the wishes of its maker . . . I consider the body of a man as being a sort of machine so built up and composed of nerves, muscles, veins, blood, and skin, that though there were no mind in it at all, it would not cease to have the same motions as at present, exception being made of those movements which are due to the direction of the will, and in consequence depend upon the mind. (Meditation Six)

2. Henryk Skolimowski, *Living Philosophy: Eco-Philosophy as a Tree of Life* (New York: Penguin/Arkana, 1992), 12.

CHAPTER 2. Why We Need to Embrace Uncertainty

1. David Lindley, *Uncertainty: Einstein, Heisenberg, Bohr, and the Struggle for the Soul of Science* (New York: Anchor Books, 2008), 19.

2. T. P. Purdy, R. W. Peterson, and C. A. Regal, "Observation of Radiation Pressure Shot Noise on a Macroscopic Object," *Science* 339, no. 6121 (2013): 801–4, doi: 10.1126/science.1231282.

3. Fred Alan Wolf, *The Dreaming Universe: A Mind-Expanding Journey into the Realm Where Psyche and Physics Meet* (New York: Touchstone, 1995), 239.

CHAPTER 4. How We Are All Connected in an Inseparable, Participatory Universe

1. Albert Einstein, Boris Podolsky, and Nathan Rosen, "Can Quantum-Mechanical Description of Physical Reality Be Considered Complete?" *Physical Review* 47, no. 10 (May 15, 1935), doi: dx.doi.org/10.1103/PhysRev.47.777.

   For a simplified explanation, see: "This Month in Physics History: Einstein and the EPR Paradox," *APS News* 14, no. 10 (November 2005): 2, aps.org/publications/apsnews/200511/history.cfm.

2. William Hermanns, *Einstein and the Poet: In Search of the Cosmic Man* (Brookline Village, MA: Branden Press, 1983), 86.

3. John Markoff, "Sorry, Einstein. Quantum Study Suggests 'Spooky Action' Is Real," *New York Times*, October 21, 2015, nyti.ms/1Gq1USJ.

4. For a detailed discussion of Bell's theorem in plain language, see Gary Felder, "Spooky Action at a Distance: An Explanation of Bell's Theorem," (online paper, 1999), felderbooks.com/papers/bell.html.

5. Markoff, "Sorry, Einstein."

6. B. Hensen et al., "Loophole-Free Bell Inequality Violation Using Electron Spins Separated by 1.3 Kilometres," *Nature* 526 (October 29, 2015): 682–86, doi: 10.1038/nature15759.

7. Vlatko Vedral, "Living in a Quantum World," *Scientific American*, June 1, 2011, bit.ly/2aikjHQ.

8. Ervin Laszlo, *The Connectivity Hypothesis: Foundations of an Integral Science of Quantum, Cosmos, Life, and Consciousness* (Albany: State University of New York Press, 2003), 11.

9. Rupert Sheldrake, *The Sense of Being Stared At: And Other Aspects of the Extended Mind* (New York: Harmony, 2003), x.

10. Skolimowski, *Living Philosophy*, 12.

11. René Descartes, *Treatise on Man* (selection from *l'homme et la formation du foetus*), ed. Claude Clerselier, trans. P. R. Sloan (Paris, 1664), bit.ly/29RFtHa.

12. Rebecca Costa, *The Watchman's Rattle: A Radical New Theory of Collapse* (New York: Vanguard Press, 2012).

13. "Remarks of President Barack Obama—State of the Union Address As Delivered" (transcript), January 13, 2016, bit.ly/1P05B5F.

14. Heraclitus, as quoted by Plato in *Cratylus*: "And here I seem to discover a delicate allusion to the flux of Heracleitus—that antediluvian philosopher who cannot walk twice in the same stream; and this flux of his may accomplish yet greater marvels . . .

Those again who read *osia* seem to have inclined to the opinion of Heracleitus, that all things flow and nothing stands." Plato, *Cratylus*, trans. Benjamin Jowett (Project Gutenberg, 1999), last updated 2013, bit.ly/2deOfWg.

15. John A. Wheeler, *At Home in the Universe* (Melville, NY: American Institute of Physics, 1997), 24.

16. Vicki Abeles, "Is the Drive for Success Making Our Children Sick?" *New York Times*, January 2, 2016, SR2, nyti.ms/1mrIjcv.

17. Jeremy Rifkin, *The Zero Marginal Cost Society: The Internet of Things, the Collaborative Commons, and the Eclipse of Capitalism* (New York: Palgrave Macmillan, 2014).

18. Timothy Morton, *The Ecological Thought* (Cambridge, MA: Harvard University Press, 2012), 17.

19. The concept of swarm theory, or swarm intelligence, was introduced by Gerardo Beni and Jing Wang in 1989, in the context of cellular robotic systems. For a good overview for the layperson, see: Peter Miller, "The Genius of Swarms," *National Geographic*, July 2007, bit.ly/1t0braz.

CHAPTER 5. Becoming the Master of Your Thinking

1. Jon Kabat-Zinn and Richard J. Davidson, eds., *The Mind's Own Physician: A Scientific Dialogue with the Dalai Lama on the Healing Power of Meditation* (Oakland, CA: New Harbinger Publications, 2013).

   His Holiness the Dalai Lama, *The Universe in a Single Atom: The Convergence of Science and Spirituality* (New York: Morgan Road Books, 2005).

   David Bohm, *Thought as a System* (London: Routledge, 1994). Much of my understanding of the nature of thought has been influenced by the pioneering work of David Bohm. Widely considered one of the most significant theoretical physicists of the twentieth century, Bohm contributed innovative and unorthodox ideas to quantum theory and the nature of thought and dialogue. A good part of what follows in this book is inspired by his insights, expanded upon and further developed by my own work and therapeutic innovations.

   Debbie Hampton, "How Your Thoughts Change Your Brain, Cells and Genes," The Huffington Post, March 23, 2016, huff.to/22FnGZP.

2. Kathy Gilsinan, "The Buddhist and the Neuroscientist: What Compassion Does to the Brain," *The Atlantic*, July 4, 2015, theatln.tc/ 1M2jWJ3.

   Research that shows positive effects on the brain chemistry of regular deep meditators is discussed in two accessible articles, with links that will take you to formal research studies: Rebecca Gladding, "Use Your Mind to Change Your Brain: This Is Your Brain on Meditation, *Psychology Today*, May 22, 2013, bit.ly/ 17Vlpmy; Brigid Schulte, "Harvard Neuroscientist: Meditation Not Only Reduces Stress, Here's How It Changes Your Brain," *Washington Post*, May 26, 2015, wapo.st/1L2DfjT.

3. Fabrizio Benedetti, Elisa Carlino, and Antonella Pollo, "How Placebos Change the Patient's Brain," *Neuropsychopharmacology* (2011): 36, 339–54, doi:10.1038/npp.2010.81 (published online June 30, 2010).

4. For detailed background on near-death experiences, see Janice Miner Holden and Bruce Greyson, *The Handbook of Near-Death Experiences: Thirty Years of Investigation*, 2nd printing ed. (New York: Praeger, 2009); Gideon Lichfield, "The Science of Near-Death Experiences: Empirically Investigating Brushes with the Afterlife," *The Atlantic*, April 2015, theatln.tc/2gRSOny.

CHAPTER 6. Moving Beyond Either-Or

1. Mark Dewolfe Howe, ed., *Holmes-Pollock Letters: The Correspondence of Mr. Justice Holmes and Sir Frederick Pollock, 1874–1932*, 2nd ed. (Cambridge, MA: Belknap Press of Harvard University Press, 1961), 109. Often misquoted as "I wouldn't give a fig for the simplicity on this side of complexity; I would give my right arm [or my life] for the simplicity on the far side of complexity" and misattributed to Oliver Wendell Holmes, Sr.

CHAPTER 7. Freeing Ourselves from the Grip of Pathology

1. Such exceptions include transpersonal therapy, which seeks to integrate spirituality and aspects of life that transcend personal identity, and family systems therapy, which works with families and couples and emphasizes the systems of interaction among family members.

2. Rebecca Orleane, *The Return of the Feminine: Honoring the Cycles of Nature* (Bloomington, IN: AuthorHouse, 2010), 104.

3. American Psychiatric Association, *Diagnostic and Statistical Manual of Mental Disorders*, 5th ed. (Washington, DC: American Psychiatric Publishing, 2013), 153.

4. Marie-Louise von Franz, *Psyche and Matter* (Boston: Shambhala Publications, 1988), 288.

5. William C. Reeves et al., "Mental Illness Surveillance among Adults in the United States," *Morbidity and Mortality Weekly Report (MMWR)* 60, no. 3, (September 2, 2011): 1–32, bit.ly/2esSTf7.

6. "Normosis," *Momento Espirita* (Brazilian radio program), August 5, 2010, bit.ly/2ajOL1L. See also P. Weil, R. Crema, and J.Y. Leloup, *Normosis: The Pathology of Normality*, 5th ed. (Petrópolis, Brazil: Editora Vozes, 2014).

7. Alfred North Whitehead, *Science and the Modern World* (1925; repr., New York: Free Press, 1997), 51.

8. Ethan Watters, *Crazy Like Us: The Globalization of the American Psyche* (New York: Free Press, 2011), 7.

9. Thomas Kuhn, *The Structure of Scientific Revolutions* (Chicago: University of Chicago Press, 1996), 89.

CHAPTER 8. From Being to Becoming: Creating Authentic Self-Esteem

1. Albert Pinkham Ryder, "Paragraphs from the Studio of a Recluse," *Broadway*, no. XIV (September 1905): 10, quoted in Barbara Novak, *Voyages of the Self: Pairs, Parallels, and Patterns in American Art and Literature* (New York: Oxford University Press, 2009): 121.

CHAPTER 9. Beyond the Mind-Body Connection

1. "What Is the Placebo Effect?" WebMD, last modified February 23, 2016, wb.md/2gdgsKa.

2. "How the Placebo Effect Enhances Healing," *Mayo Clinic Health Letter*, April 23, 2014, mayocl.in/29qw5gH.

3. Maryann Mott, "Did Animals Sense Tsunami Was Coming?" *National Geographic News*, January 4, 2005, bit.ly/29nfADC.

4. Ibid.

5. Allan Combs and Mark Holland, *Synchronicity: Through the Eyes of Science, Myth and the Trickster*, 3rd ed. (New York: Da Capo, 2000).

6. Carl Jung, *Synchronicity: An Acausal Connecting Principle* (extract from vol. 8. of *The Collected Works of C. G. Jung*), trans. R. F. C. Hull (1960; repr., Princeton, NJ: Princeton University Press, 2010).

CHAPTER 10. Entanglement Is the Heartbeat of Love

1. Oscar Wilde, *The Importance of Being Earnest: A Trivial Comedy for Serious People* (1915; Project Gutenberg, 2006), gutenberg.org/ebooks/844.

CHAPTER 11. Coherent Communication

1. Abraham Lincoln, "Address at a Sanitary Fair, Baltimore, Maryland," April 18, 1864.
2. William Isaacs, *Dialogue and the Art of Thinking Together* (New York: Crown, 1999).
3. David Bohm, *On Dialogue* (New York: Routledge, 1996).
4. Natalie S. Bober, *Thomas Jefferson: Draftsman of a Nation* (Charlottesville: University of Virginia Press, 2007), 285.

CHAPTER 12. Not Just Semantics

1. D. David Bourland and Paul Dennithorne Johnston, *To Be or Not: An E-Prime Anthology* (Forest Hills, NY: International Society for General Semantics, 1991), xii.
2. George Lakoff, "Understanding Trump," The Huffington Post, updated July 24, 2016, huff.to/2ag7JY6.
3. For a good selection of Whorf's writings, see John B. Carroll, ed., *Language, Thought, and Reality: Selected Writings of Benjamin Lee Whorf* (Cambridge, MA: MIT Press, 1964).
4. Alfred Korzybski, *Science and Sanity: An Introduction to Non-Aristotelian Systems and General Semantics*, 5th ed. (Forest Hills, NY: International Society for General Semantics, 1995).
5. Plato, *Cratylus*, trans. Benjamin Jowett, bit.ly/2deOfWg.

INDEX

intention, 164–65

interactive universe, 15–16

interconnection, 31–44, 113–14. *See also* connectedness; inseparability
  big picture, 35–39
  entanglement, 32–35
  fragmented thinking and, 36–37, 103
  physics experiments on, 31–33
  visualization on, 42
  whole as more than sum of parts, 36

intimacy, 136, 137

intuition, 104, 105–7
  reintegrating with intellect, 106–7

is (word use), 152, 153, 155

Isaacs, William, 138

Jefferson, Thomas, 138

Johnston, Paul, 151

Jung, Carl, 110

Kepler, Johannes, 5

knowing, 104–6

Korzbyski, Alfred, 152

Kuhn, Thomas, 84

Lakoff, George, 151

language, 151–61. *See also* semantics
  dictionaries, 156
  E-Prime, 152, 157–61
  worldview and, 151–52, 158–61

Laszlo, Ervin, 34

law/laws of the universe, 165–66

learning, 95, 96

light (photons), 31–33
  wave-particle duality, 22, 63, 155

Lincoln, Abraham, 135

listening, 140–42

literal thought, 55–57, 59, 62, 78–79
  to be verbs and, 152

local reality, 32–33

locality, 33

love, 115–29. *See also* relationships
  autonomous, dependent, and
    independent individuals, 116–17
  clarifying word meanings, 136
  falling in love, 115–16
  personalization of issues, 119–21
  relationship problems, 117–21
  shifting the energy of relationships, 121–22
  uncertainty and romance, 127–28

mechanistic paradigm/worldview, 6–8, 10, 22, 30, 36
  language of, 47–48, 151–53, 161
  perfection, seeking, 97

mechanomorphism, 48

medical field, 37

medical specialization, 37

meditation, 46, 102–3, 172n2

memories, 27–28

men/masculine qualities, 104, 106

mental illness, 75–77

midlife crisis, 99

mind-body connection, 101–14
  brainwashing and, 102–3
  false reality of, 101, 102
  harm from divisions created, 104–6
  inseparability and, 101
  intellect and intuition, 103–7
  problems with, 101–14
  sixth sense and, 108–9
  synchronicity and, 110

misplaced concreteness, 79, 159

mistakes, 94–96, 160

Morton, Timothy, 43

motion, Newton's first law of, 19–20, 30

near-death experiences, 47

neuroplasticity, 46–47, 65

Newton, Isaac, 6, 13, 15, 19, 32, 33, 36, 61

nonlocality, 32

possibilities, 3–4. *See also* potential
  potentiality and, 8, 10, 30
  space between thoughts and, 54–55
  uncertainty and, 8, 16, 20, 60
  wave collapse and, 27
potential, 10, 21–30, 125, 153
  beliefs and, 23–24, 25–27, 28
  for infants, 23
  pure, 25–27
  quantum physics concepts, 21–25,
    27
  recovering, 16, 21–30
  releasing your past, 27–30
  space between thoughts and, 54–55
  wave collapses and, 21, 22–25,
    28–30, 125
potentiality, 8, 9, 10, 30
prayer, 34
predictability. *See* certainty
present, being fully in, 97–98, 143
probability waves, 22
psi phenomena, 34
psychology, 71–84
  coping mechanisms, 88–91
  diagnosis and, 74–75
  diagnosis disorder, 77–82
  Diagnostic Manual (DSM), 74, 79
  prevalence of problems, 75–77
  traditional, 74–75
psychosomatic (terminology), 102
psychotherapy
  diagnosis, place for, 82–84
  medicating patients, 81
  myth of objectivity and, 73–75
  need to address feelings, 81–82
  objectivity and, 71, 172n1
  pathologizing and, 73–75
  reiteration of past experiences in,
    75, 76
  subjectivity and, 74–75
  suffering, epidemic of, 75–77, 84
  traditional, 74–75
pure potential, 25–27. *See also* potential

quantum physics, 3–4, 8–10, 20
  change and, 9, 10
  entanglement, 32–35
  inseparability and, 9, 43
  "Living in a Quantum World", 34
  potentiality and, 8, 9, 10
  primary principles, 8–9, 12
  superposition, 22, 153
  uncertainty and, 8, 14–16
  wave collapse, 21, 22–25
  wave-particle duality, 22, 63, 155
quantum worldview, 5–10, 20, 22
  applications of, 2–4
  Capra on, 1–2
  change and, 20
  communication and, 131
  inseparability and, 9
  participating in, 1–2
  personal transformation and, 2–3
  possibilities and, 3–4
  potentiality and, 8, 9, 10, 27
  reality in, 27, 31
  relationships in, 127–28
  uncertainty and, 8, 14–16, 73
questions
  on communication, 144–45
  either-or, 63
  of identity, 18–19
  new/perpetual, 65, 66–67
  "What if?", 98–100
  "Who am I?", 86–88, 158
  "Who will I be?", 18–19

reactions, observing vs. becoming,
    142–45
reality, 20, 31, 63–64, 85
  acausal, 32–33
  entanglement and, 35
  language and, 151–52, 157
  local reality, 32–33
  objective, 10, 32, 57, 59, 73
  participatory, 64
  potential and, 27
  in quantum universe, 27, 31
  subjectivism and, 64
  thoughts and, 25–26, 27, 161

ABOUT THE AUTHOR

Psychotherapist, couples counselor, and speaker Mel Schwartz is an emerging voice in the field of personal transformation. He is one of the first contemporary practicing psychotherapists to distill the basic premises of quantum theory into therapeutic approaches that enable people to overcome their challenges and live to their fullest potential. Mel is an example of the self-actualization techniques he unveils. While in his early forties, with two young sons, and in the middle of a comfortable life as an entrepreneur, he envisioned a life with a much deeper purpose. He decided to follow his calling and move in an entirely new direction, one filled with uncertainty. *The Possibility Principle* is one of the results of that transition.

He is also the author of *The Art of Intimacy, The Pleasure of Passion*. He has written more than one hundred articles—read by more than 1.5 million readers—for *Psychology Today* and his blog, *A Shift of Mind*. In private practice for more than two decades in Manhattan and Westport, Connecticut, Mel also works with clients globally by Skype. He earned his graduate degree from Columbia University and has been a keynote speaker at Yale University. To read about Mel's workshops, talks, and offerings, please go to melschwartz.com.

ABOUT SOUNDS TRUE

Sounds True is a multimedia publisher whose mission is to inspire and support personal transformation and spiritual awakening. Founded in 1985 and located in Boulder, Colorado, we work with many of the leading spiritual teachers, thinkers, healers, and visionary artists of our time. We strive with every title to preserve the essential "living wisdom" of the author or artist. It is our goal to create products that not only provide information to a reader or listener, but that also embody the quality of a wisdom transmission.

For those seeking genuine transformation, Sounds True is your trusted partner. At SoundsTrue.com you will find a wealth of free resources to support your journey, including exclusive weekly audio interviews, free downloads, interactive learning tools, and other special savings on all our titles.

To learn more, please visit SoundsTrue.com/freegifts or call us toll-free at 800.333.9185.